THE BASIC RECORDING MANUAL

By

Robert Dennis

and

Daniel Dennis

ORIGINAL TEXT BY
ROBERT DENNIS

COMPUTER RENDITION, ADDITIONAL
DRAWINGS, UPDATES TO TEXT BY
DANIEL DENNIS

PUBLISHED BY:

14611 EAST NINE MILE ROAD
EASTPOINTE, MI 48021 USA

Originally published as
Multitrack Recording One Lesson Text
Original Text: © 1996 Robert Dennis.
Chapter 8: © 1997 Robert Dennis.
Revisions: © 1999 Robert Dennis.
Re-Revisions: © 2000, 2001,
Renamed
Basic Recording Manual
© 2005 Robert Dennis
Re-Revisions: © 2021, Daniel Denis.

Published by (with permission of the copyright holder):
Recording Institute of Detroit, Inc.
14611 E. Nine Mile Road
Eastpointe, MI 48021

FIRST PRINTING: 1996
SECOND PRINTING: 1997
THIRD PRINTING: 1999
FOURTH PRINTING: 2000
FIFTH PRINTING: 2001
SIXTH PRINTING: 2005
SEVENTH PRINTING: 2021

All rights reserved.

No part of this book may be reproduced, stored in a retrieval system, or transmitted by any means without written permission of the copyright holder.

Neither the author nor the publisher assume any liability whatsoever because of misinformation or inaccuracies in this text and give no warrantees of any kind. The author and publisher assume no liability of any kind, including but not limited to loss or injury, direct, indirect or consequential.

The purpose of any part of this text is advice to the reader in the operation of the equipment referred to in this text. It is not intended to replace any manual published by the manufacturer of any equipment, or any other licensed publication.

BASIC RECORDING MANUAL

TABLE OF CONTENTS

TITLE PAGE		I
COPYRIGHT AND LEGAL NOTICES		II
TABLE OF CONTENTS		III
AUTHOR'S PREFACE		IV
CHAPTER 1	THE FIRST BASIC	5
CHAPTER 2	SOUND	9
CHAPTER 3	USING MICROPHONES	15
CHAPTER 4	MORE ABOUT MICROPHONES	23
CHAPTER 5	USING A BASIC RECORDING SYSTEM	31
CHAPTER 6	MIXDOWN	39
CHAPTER 7	RECORDING RAP AND HIP-HOP	47
CHAPTER 8	USING DIGITAL SYSTEMS	51
CONCLUSION		55

AUTHOR'S PREFACE 1995

Introduction

I've been teaching multitrack recording techniques for over 30 years. The demand to learn recording quickly is increasing as the subject is becoming more complex. At the Recording Institute of Detroit, we have increased training time in our professional program from 80 clock hours to over 300 hours. The person trying to "get started" or "hobby" with recording has been left behind.

Some little time ago, I decided that you could teach someone enough about recording in one day to get them started. This short text is the text of our "One Day Recording Workshop" that we introduced. With this "cram" course one can learn some important basics about recording and let experience be an additional teacher. After getting some experience, it is natural that a more in-depth study will take the student to the next level.

If reading and using the information in this text "peaks" your interest in recording, Recording Institute of Detroit still offers its professional program and seven published textbooks on recording.

Robert Dennis, 1995

COAUTHOR'S PREFACE 2021

Having taken over the training of engineers in place of my late father, I can attest to the veracity (truthfulness) of his statement from over 25 years ago that the field is only getting more complex.

The Hobbyist is alive and well and to be found on Youtube and other social media locations. But a true understanding of recording and mixing does take much more extensive training. This text can still be used to get someone started.

This text is used at the very beginning of the professional training given by the Recording Institute of Detroit. You might wish to go much farther than the basic recording explained herein.

But you have to start somewhere, and the very basics is a good place.

Daniel Dennis, 2021

CHAPTER 1 – THE FIRST BASIC

Multitrack Recording

Multitrack recording is a process where different signals are recorded separately, but in time with each other, such that they can be combined later.

This differs from live "direct to tape" in which the individual sounds are already blended into a signal such as what you would find on a music CD or MP3.

The advantage to multitrack recording is the ability to replace, edit, remove or add performances to get the best possible product.

To make a song into a product that you would put onto a CD, there are various stages.

1. Recording Basic Tracks: To record correctly all usable tracks as a basis for the song.

   In a live recording, the concentration should be on any tracks that should not be overdubbed, such as drums, and the performance of any musician who will not be returning (out-of-town musicians, studio musicians, etc.

2. Overdubbing: To replace any imperfect performances and to add all performances needed to complete the production.

3. Mixdown: To get the sound of the production to the best possible balance of all instruments and performances for that song.

4. Mastering: To take all of the separate mixdowns and, by processing and editing, to get them to sound like they belong together on the same album (similar levels, etc).

Each step will only be as successful as the step before.

Please note that recording correctly is the first step. And when talking about recording there is one thing that is more important than anything else.

This primary concern is summed up in one phrase: Signal-to-Noise Ratio. In simplest terms, signal would mean whatever you are trying to record. Noise would be anything you are not trying to record.

A very simple example would be two singers being recorded to two separate tracks using two separate microphones.

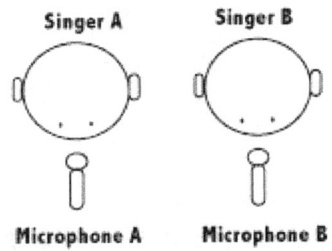

Two Singers

Singer A, being recorded to track 1, would be considered "signal", but only for track 1.

Singer B, even though they are being recorded at the same time to a different track would be considered to be "noise" as far as track 1 is concerned.

Likewise, as far as track two is concerned, Singer A would be considered to be "noise" and Singer B would be "signal."

A lot of what an engineer does in a recording session comes down to this principle.

How the room is set up, how microphones are placed, how the knobs and controls of the mixing board are adjusted; all of these things are dictated by this one principle: Signal-to-Noise Ratio.

Level Basics

The Decibel

An understanding of level starts with an understanding of the term "decibel."

Decibel does NOT mean volume. This is a common misunderstanding by those who have never actually had the term defined properly.

The decibel (abbreviated "dB") is actually a measurement of change.

0 dB is a reference level. In some instances 0 dB does mean "no audible sound." In other instances 0 dB means the loudest or highest possible level. 0 dB is any level or volume that you are comparing another level or volume to.

A change of 1 dB is an increase or decrease of about 12%. In practical terms, when dealing with relative volume a decibel it is the smallest amount of change that can be heard under perfect circumstances. It is not necessarily noticeable.

A 3 dB change is a noticeable change. Anything that is 3 dB louder or quieter is noticeably louder or quieter. Though noticeable, it is still noted that 3 dB is a *small* change.

A 6 dB change is, in effect, a doubling or halving of the volume. A level at +6 dB is twice as loud as your reference level of 0 dB. A level of -6 dB would be half as loud as your reference volume at 0 dB.

It should also be noted that 6 dB is *practically* twice the volume. In truth it is closer to 1.995 times the volume. However, this small difference is negligible and can be ignored.

Chapter 1 – The First Basic

A 20 dB change is a change of either exactly a tenth or exactly ten times the reference volume. +20 dB is 10 times as loud as your reference volume of 0 dB. -20 dB is one tenth as loud as your reference volume of 0 dB.

Human Hearing and the Decibel Scale

As in many aspects of human perception, the human ear hears changes in terms of percentages. Humans perceive change it terms of "how much is it different?" rather than "how many is it different?"

As a result, like many other scales the decibel scale is logarithmic in nature.

In simplest (very non-mathematical) terms, this means that as one side of the scale adds, the other side multiplies.

A very easy example of what a logarithmic scale is could be gambling at a roulette table. If you bet black/red, you get double your money if you are right.

Let's take an example of guessing right six times in a row, betting all winnings each time.

Each time you add one spin, you multiply your money by 2 times.

Your money will increase, starting with $1, on a logarithmic scale as follows:
- Spin 1 - you win $2
- Spin 2 - you win $4
- Spin 3 - you win $8
- Spin 4 - you win $16
- Spin 5 - you win $32
- Spin 6 - you win $64

So, in terms of decibels and volume, +6 dB is 2 times as loud, +12 dB is 4 times as loud, +18 dB is 8 times as loud and so on.

To the human ear, the difference between a 100-watt amp's dB output and a 200-watt amp's dB output would be the same as the difference between the outputs of a 600-watt amplifier and a 1200-watt amplifier (2 times the power output).

This is a chart comparing different dB levels and the amount of change in volume:

Decibel Changes vs. Volume

Students having access to PAS Demos, listen to PAS Demo 6, Level Changes
Signal-to-Noise Ratio and Different Formats

The actual definition of Signal-to-Noise (abbreviated S/N) ratio is: The dB difference between the signal and noise.

This can be external noise (such as leakage from other sound sources) and/or noise generated by the process (such as the use of analog tape).

Below is a chart with some examples of Signal-to-Noise Ratios for different recording formats and devices.

Realize that these are approximate levels. They assume a professional level of recording, getting the best possible signal and the least possible noise:

Format or Device	S/N Ratio	Approximate Volume Difference between Signal and Noise
Analog Cassette Tape Recording	48 dB	250X
Professional Level Analog Tape Recording, Phonograph	55 - 70 dB	560X - 3150X
CD Recording (16 Bit), Professional Analog Console	92 - 96 dB	40,000X - 65,000X
ADAT Digital Multitrack Tape Recording (20 Bit)	120 dB	1,000,000X
Full DVD Quality	144 dB	16,000,000X

Chart of Comparative Signal-to-Noise Ratios for Different Formats

The significance of the last example should not be glossed over. On the dB SPL (Sound Pressure Level) scale, the difference between the threshold of hearing (0 dB-SPL) and the Threshold of Pain (140 dB-SPL) where half the people feel physical pain listening to something so loud, is *smaller* than the signal-to-noise ratio of properly recorded DVD quality audio.

CHAPTER 2 – SOUND

Wave Shape

Imagine an aquarium that you could keep pet fish in.

The sides are glass and its about 4 feet wide, two feet deep and about 3 feet tall. If you walked over to this aquarium and wiggled your finger in the water, you would see waves appear in the water.

The waves would consist of build-ups of extra water next to valleys of less water and these waves would move out from the disturbance of your wiggling finger in all directions.

Sound travels in a similar manner, except the medium is air rather than water.

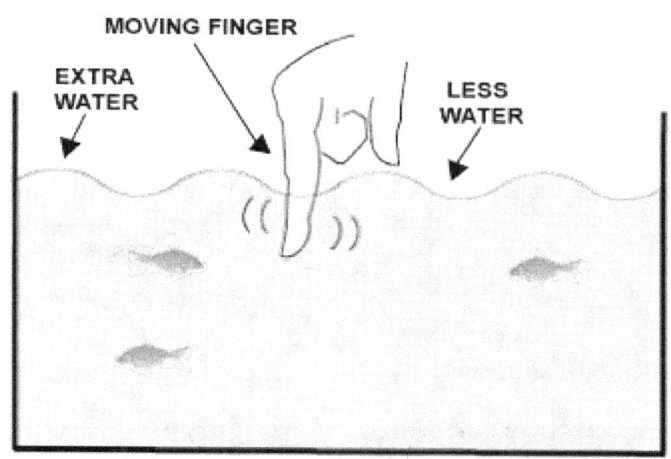

Waves Caused by Disturbing Water

Definition of Sound

To have sound, you have to have something vibrating between 20 times a second and twenty thousand times a second. The vibration may come from a guitar string, a drumhead or a vocal cord; but to get sound, you have to have something vibrating.

The vibrating source causes waves of air pressure changes.

In the water wave, there is more water in the area of the buildup and less water in the valleys in between the build-ups. In the areas of the build-up, there is more water pressure, because there is more water.

In the sound wave of air, the areas of build-up have more air particles and there are less air particles in the valleys between the build-ups.

As the vibrating source moves back and forth, it first compresses the air particles together and then spreads them apart. In the areas of buildup, there is more air pressure and there is less air pressure in the valleys.

The vibrating source causes a wave of changing air pressure and is called a Sound Pressure Wave.

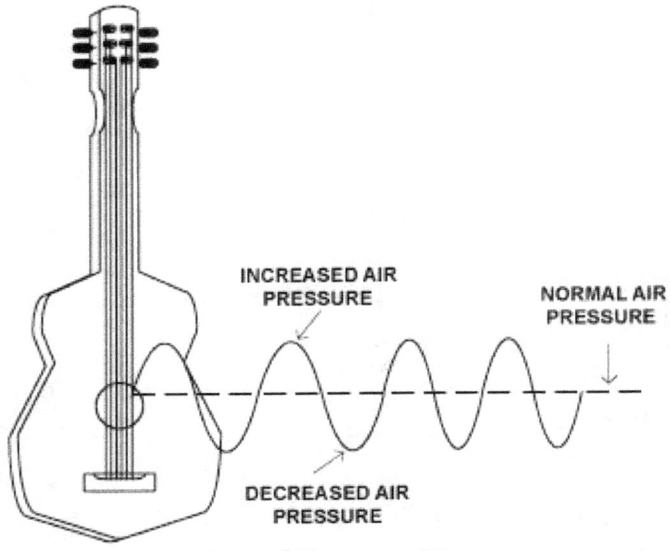

Sound Pressure Wave

The definition of sound is:

A MOVING AIR PRESSURE WAVE CAUSED BY A SOMETHING VIBRATING BETWEEN 20 TIMES A SECOND AND 20,000 TIMES A SECOND.

As the changing air pressure of the sound pressure wave moves past a person's ear, the changing pressure causes the ear drum to vibrate in and out, much the same way the vibrating source moved.

Our hearing mechanism causes the vibration of the eardrum to become sound to the brain.

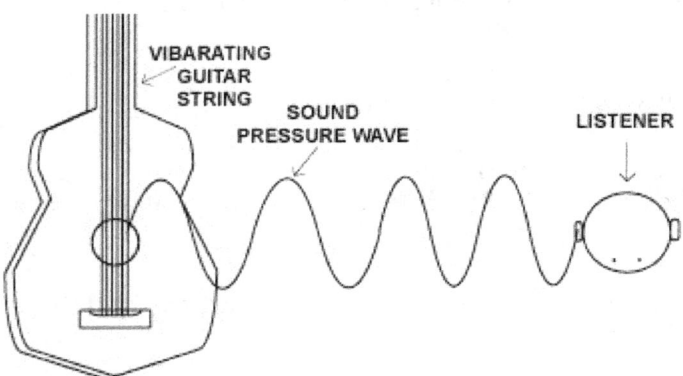

Eardrum Responding to Sound Pressure Wave

Sound Absorption and Reflection

You have probably noticed that your voice sounds different in different kinds of rooms.

In an empty room, your voice sounds hollow. Once you move in the furniture and drapes, the hollow sound disappears. The difference is caused by the amount of sound absorption and reflection.

The sound pressure wave will reflect off of hard surfaces. The sound pressure wave is absorbed by soft porous surfaces.

In a typical living room, much of the sound is absorbed by the soft-porous surfaces of the carpeting, furniture and drapes. When a room is empty, there are a lot of hard walls exposed for the sound pressure waves to reflect off.

In a typical bathroom, there are a lot of hard surfaces that aren't covered up. When you sing in the shower, your voice lingers on.

This continuing of the voice is called reverberation.

Mr. Baffle Demonstration

In most studios, there are sound absorbing baffles in use. The baffles in many professional baffles are filled with very dense fiberglass. Fiberglass absorbs sound.

The outer shells of the baffles are made of steel. One side has holes, which allows the sound to enter the baffle where it will be absorbed by the inner fiberglass.

The other side is hard and smooth, with no holes, and will reflect the sound.

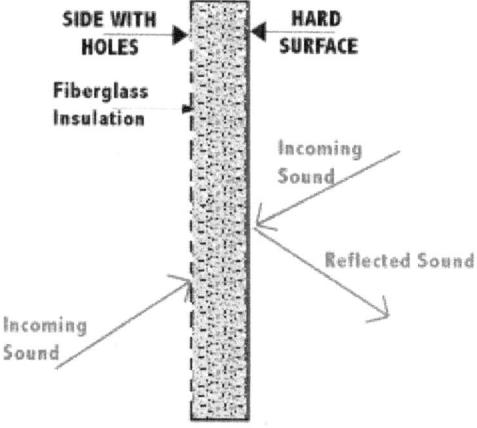

Sound Isolation Baffle

The instructor will have you talk into both sides of the baffle.

Talking into the absorbent side of the baffle will result in your voice being "sucked up" by the baffle. Talking into the reflective side of the baffle will result in your voice being reflected back to you.

As you hear others talk into the baffle, you will notice that their voices are much louder in the room when they talk into reflective side of the baffle.

You will notice that their voices are much softer in the room when they talk into the absorbent side.

Alternately, this demonstration can be done with foam insulation and a hard, flat surface (preferably metal, glass, or a whiteboard).

Sound Isolation

Isolation

In multitrack recording, we are recording separate recordings that are in time with each other. Each separate recording is recorded onto a separate track.

To record the basic instruments for a live band, you would most likely record the following tracks:

Track 1 Bass Guitar
Track 2 Foot (Bass Drum)
Track 3 Snare Drum
Track 4 High Hat (double cymbals)
Track 5 Overhead Drums (Left Channel)
Track 6 Overhead Drums (Right Channel)
Track 7 Rhythm Electric Guitar #1
Track 8 Rhythm Electric Guitar #2
Track 9 Lead Electric Guitar
Track 10 Vocal

Additional tracks would be used for other parts like additional guitars and background vocals.

Often the vocal will perform the tune two or three times on different tracks and the best performance will be used in the final product. Alternately, in more modern usage a vocal track is copied and the copy has a tuning function applied to it.

The basis of multitrack recording is that you can control the sound of different tracks in the final product.

When blending the different tracks together into the final sound (the mixdown process), you can control the volume and tone of each of these tracks. You can decide that a guitar or bass line needs to be done again, throwing away the first recording and using the later recording.

To achieve the purpose of multitrack recording, the sound that is recorded on each track must be kept separate. If the guitar can be heard in the drum tracks, for instance, the guitar part cannot be replaced with a new performance because the old guitar will still be heard through the drum tracks.

The process of keeping the sounds separate is called Isolation.

The term leakage is used if the sound of an instrument gets into another instrument's microphone.

Small rooms off the main studio with glass windows are called Isolation Rooms. They are also called Isolation Booths or Sub-Rooms.

These isolation rooms can be used for instruments that are very loud, like electric guitars. Since the amplifier is in a separate room, its loud sounds don't leak into the microphones in the main studio.

The isolation rooms can also be used for low-level instruments, such as acoustic guitars. Putting the softer instruments in the isolation room prevents the sound of instruments in the main studio from getting into the soft instrument's microphone.

Chapter 2 – Sound

Placing Baffles for Isolation

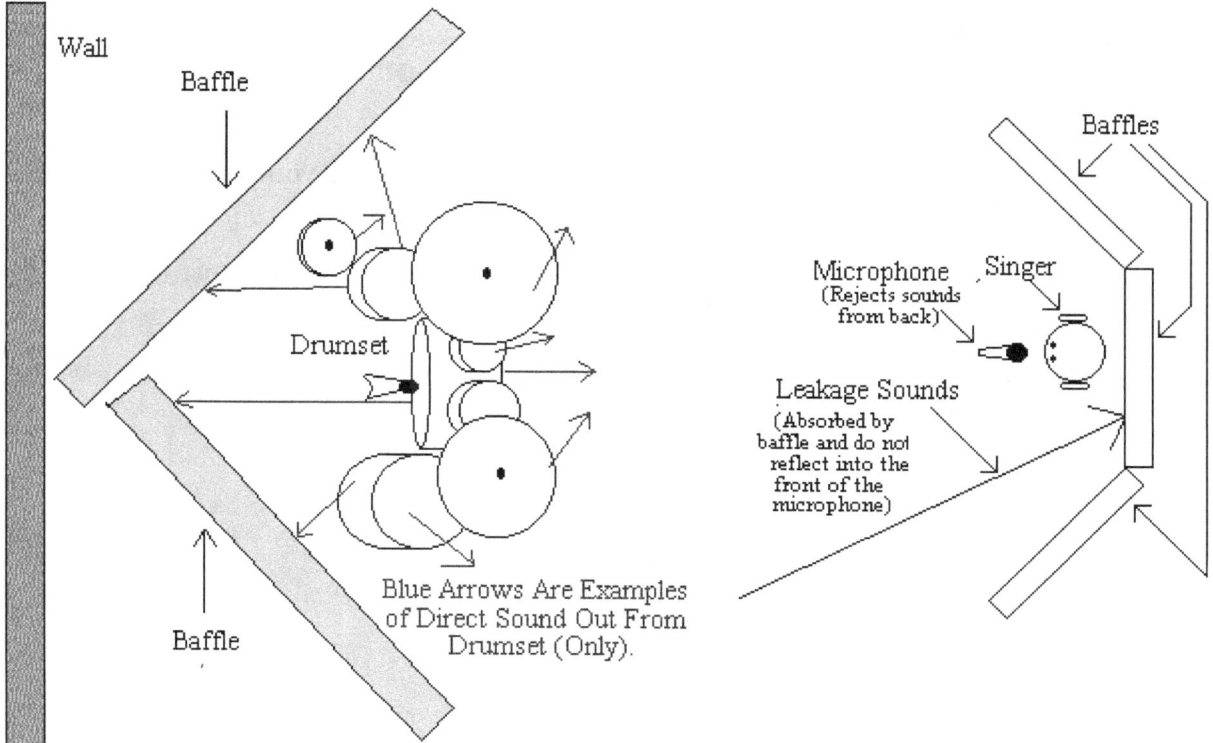

Placing Baffles for Isolation

The biggest concern when recording multiple sources is not the direct sound as most microphones have a directional aspect (explained in the next chapter). Baffles are usually placed *behind* the instruments to achieve isolation.

The microphones that are used are usually directional microphones.

The front of the microphone picks up the sound very well and the back of the microphone will not pick up sound very well.

Microphones placed on other instruments in the room are placed so that the back of the microphone points towards the loud sound source (like the drums).

If an instrument is in front of a hard wall, the sound from the instrument will reflect off of the wall into the room.

This is very noticeable with a drum kit.

Putting baffles behind the drum kit means that this reflection is eliminated and the drums will not project into the studio as much.

The drum sound in the room is about half-volume with baffles placed behind the drums.

As an example, let's say that you were using the main studio to record the drums and vocal.

The guitar amplifiers are in the isolation rooms and all you have to worry about is that the drums don't leak too badly into the vocalist's microphone.

You would place the vocalist and drums facing each other. The eye contact between them is important for them performing their parts together. Behind the vocalist and behind the drums, you would place sound absorbing baffles.

You would place the vocalist's microphone very close to his/her mouth and you would point the back of the microphone at the drums.

The drum sound projecting from the drums to the vocal area is cut about 50% because of the baffles behind the drums. The back of the microphone tends to reject this sound.

The sound that does get through is absorbed by the baffle behind the vocalist, and does not get reflected back into the front of the microphone.

CHAPTER 3 – USING MICROPHONES

Definition / Purpose of Microphones

A microphone is a device that changes the sound into electrical energy.

The sound wave that we described in the last chapter is referred to as acoustic energy. The microphone changes this to electrical energy that can run through wires to a different location. The electrical energy made from the sound wave is called an audio signal.

To understand electrical energy you need to know about the building block of physical substance: the atom. The smallest particle of anything physical is the atom. This particle is so small that millions of them are in the head of a pin. If you could see an atom, it would look like a sun with planets revolving around it. The center of the atom (the "sun") is called a nucleus. The other particles revolving around it (the "planets") are called electrons. Electrical current is the flow of electrons away from their home atoms and down a substance.

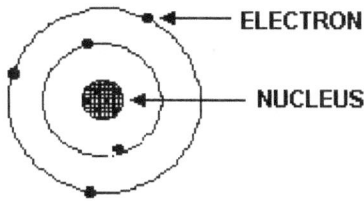

Parts of an Atom

Some substances have atoms that hold onto the electrons very loosely. In these substances, called conductors, electricity can flow easily. Most of the metals, including silver and copper, are examples of conductors. Atoms in other substances hold onto their electrons very tightly. In these substances, called insulators, little (or no) electricity can flow. Glass and plastic are examples of insulators. Non-purified water is a conductor and air is an insulator.

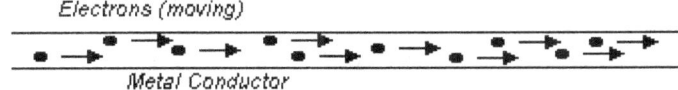

Electrons Moving in a Conductor

Electricity flows in circuits. A good example of a circuit is in the common flashlight. A wire comes from the positive side of the battery, through a switch to a light bulb. The screw base of the light bulb is connected to the negative side of the battery. When the switch is turned on, the electrons come out of the negative side of the battery, through the light bulb, through the switch and back to the positive side of the battery. Thus, the electrons have a complete circular path (circuit) from the negative side of the battery back to the positive side of the battery.

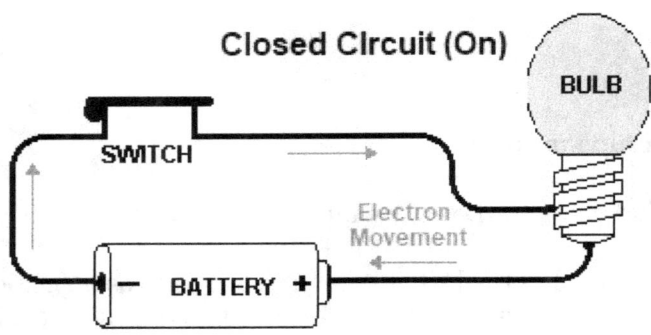

Light Switch Turned On

When the switch is turned off, electricity cannot flow through it and the electricity will not flow through the light bulb.

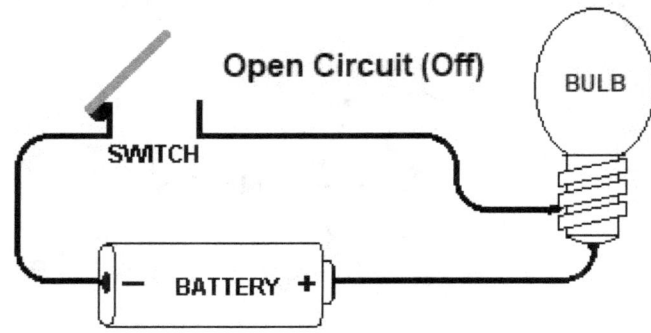

Light Switch Turned Off

The microphone has a diaphragm. This diaphragm is similar to the eardrum of a person. When the sound wave goes past the diaphragm of the microphone, the diaphragm will wind up vibrating in and out. The inward motion of diaphragm causes electrons to flow in one direction and the outward movement of the diaphragm cause the electrons to flow in the opposite direction. The audio signal out of the microphone has the electrons flowing in one direction, and then reversing direction.

The purpose of the microphone is to pick up the sound of the instrument and convert it into and audio signal. The audio signal is then run though cables into the control room, where it can be adjusted by the recording interface and sent to the recording program.

Chapter 3 – Using Microphones

Microphone Types

There are three basic types of microphones. The type of microphone has to do with the construction of the microphone, and what has to move in the microphone, to produce the audio signal.

The Ribbon microphone consists of a thin ribbon that is suspended between the poles of a magnet. A magnet has two poles (a North Pole and a South Pole). Between these "poles" there is an invisible magnetic force that will attract iron or steel. Like poles (two North poles for instance) will repel but two opposite poles (North and South) will attract. A conductor moving in a magnetic field will cause the electrons in the conductor to move. Thus, the movement of a conductor in a magnetic field causes electricity in the conductor.

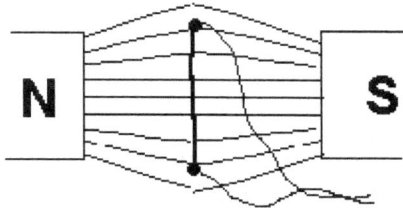

Ribbon Microphone

The ribbon in the Ribbon Microphone acts as a diaphragm. As the sound wave passes the ribbon, the ribbon will move back and forth in the magnetic field. Thus, the sound wave causes electricity in the ribbon, creating an audio signal.

The Dynamic Microphone consists of a coil of wire that moves in a magnetic field. The Dynamic Microphone has a diaphragm (usually plastic) that moves because the sound wave passes over it. The coil of wire is attached to the diaphragm so that the coil will move in the magnetic field as the sound wave passes over the diaphragm.

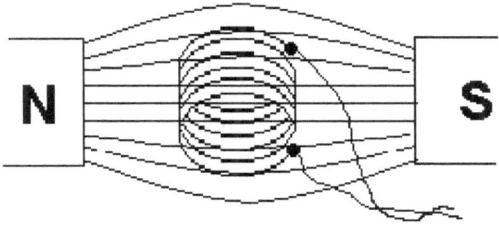

Dynamic Microphone

The Condenser Microphone has a condenser in it. A condenser is made up of two conductive plates that are close together but not touching. In a condenser microphone, one of the plates of the condenser is a diaphragm that moves with the sound wave. The other plate is fixed. This means that the one plate gets closer to a further from the other plate as the diaphragm moves. This causes a change in voltage across the plates.

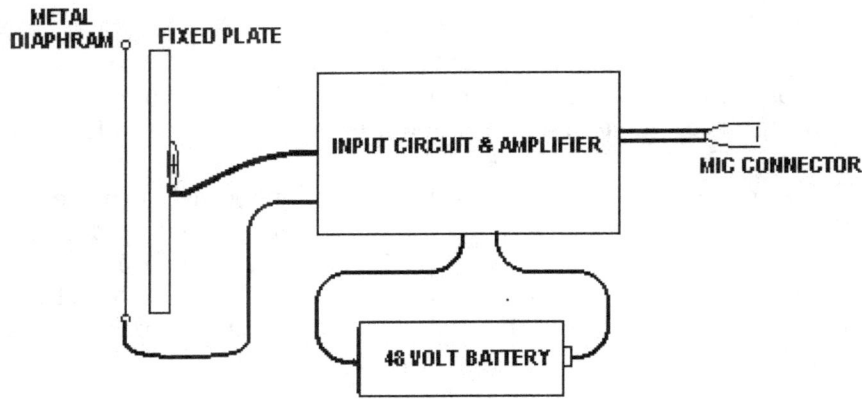

Condenser Microphone Wiring

In the condenser, microphone's case is a small (usually transistor) amplifier. The condenser is part of a circuit that feeds the input of the amplifier. Because of this circuit, the diaphragm movement causes a changing voltage at the input of the amplifier. Thus, the output of the amplifier (the output of the microphone) has an electrical voltage that changes as the diaphragm moves. Thus the diaphragm movement causes an audio signal out of the microphone.

A battery or power supply of 24 to 48 volts is required to power the amplifier and the circuit that feeds the amplifier. Although many condenser microphones can be powered from a battery, they are usually powered by a 48-volt supply in the recording interface. The same microphone cable that is used to get the audio signal to the recording interface is used to get the 48 volts from the interface to the condenser microphones. Since the 48-volt power supply is invisible, the system is called phantom powering.

Choosing the Type of Microphone

Each of the microphone types have different characteristics which make them pick up the sound of the instruments differently.

The output level of a microphone has to do with how strong the audio signal is out of the microphone. A microphone with good output level will have a stronger audio signal out of the microphone. Output level is important for low-volume instruments such as acoustic guitar. Output level would be important if you were micing medium-level instruments (such as vocals) from far away.

Because of the amplifier in the microphone's case, the condenser microphone has the highest output of all microphone types. Thus the condenser microphone would be usually be used for things like acoustic guitars and micing instruments from a distance.

The ruggedness of a microphone has to do with how much rough use a microphone can withstand and still function properly. Ruggedness is very important for live-sound mixing where the microphones have to be put up and down constantly to do different shows. The musicians on stage often move around during a performance meaning that microphones can be roughly treated while they are in use.

The dynamic microphone is, by far, the most rugged type of microphone. This is because the other two types of microphones (condenser and ribbon) are fragile. The ribbon is a thin strip that has to constantly flex and therefore will eventually break. The condenser diaphragm is close to the other condenser plate and cannot touch it; the condenser also has fragile circuitry in the case of the microphone.

Chapter 3 – Using Microphones

The dynamic microphone is the microphone of choice for almost all live-sound work.

The quickness of the diaphragm has to do with how fast it will respond to the attack of the instrument's sound. Some instruments (like drums and cymbals) are struck to be sounded and, therefore, have a very quick attack.

Both the condenser and ribbon microphone's diaphragm can move very quickly. This is because of how slowly the diagram moves in a dynamic microphone. The dynamic microphone's diaphragm is slow because it is attached to a coil of wire. The coil weighs much more than the diaphragm and, consequently, slows down the diaphragm's movement.

The ribbon and condenser microphones are used for things like cymbals. These microphones are also used for lead vocals because the diagram movements of these microphones best match the changes in the sound wave and because the vocal is often the most critical track for sound quality.

The distance that the diaphragm can move freely is an important characteristic for loud bass sounds such as rock and roll bass guitar and lower-frequency drums (toms and foot drum). The high bass energy requires the diaphragm too move a greater distance.

The dynamic microphone's diaphragm can move the greatest distance. The condenser microphone's diaphragm, with its very closely spaced plates can move the least distance.

Because of this, the dynamic microphone is the microphone of choice for loud bass guitar and for foot drums (and tom-toms).

The Different Microphones Demonstration

Students having access to PAS Demos, Listen to Demo 1.

Your instructor will play an audio demo made with high-quality studio microphones. You will hear a foot drum with a condenser microphone and then a foot drum with a dynamic microphone. The two sounds will then be repeated and butted right together so that you can best hear the difference. The condenser microphone will make the foot drum sound like a big Quaker oats (cardboard) box and the dynamic microphone will sound much more "real."

After the foot drum demonstration you will hear a high-hat cymbal recorded with two types of microphones. The cymbal recorded with the dynamic microphone will sound more dull. The cymbal recorded with the condenser microphone will sound much more "real."

Microphone Patterns

The Microphone pattern shows the direction that the microphone will pick up in. There are four common patterns explained below. Each microphone will have a specific pattern, which is achieved by how the diaphragm is encased. It does not matter what type of construction the microphone pattern has (ribbon, dynamic or condenser); directional patterns can be given to any type of microphone.

The cardioid microphone pattern has the most pickup from the front of the microphone diaphragm. There is less pickup from the sides (about half strength} and the least pickup from the back of the diaphragm (about 1/10 the signal strength). The cardioid microphone is designated with a heart. The cardioid microphone pattern is the most-used pattern. Sometimes it is referred to as an uni-directional microphone (meaning "one" direction).

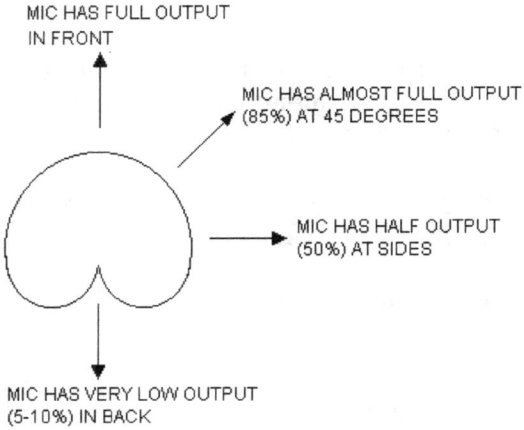

Cardioid Microphone Pattern

The omni-directional microphone pattern gives the microphone the same pickup from all angles. This pattern is designated with a circle.

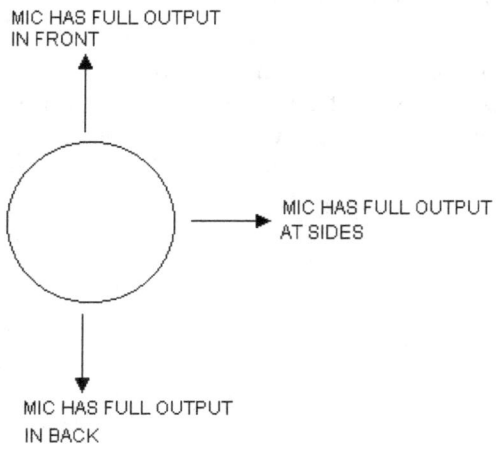

Omni-Directional Pattern

The bi-directional microphone pattern gives the microphone maximum pickup from the directly in front and directly behind the diaphragm. There is no pickup from the direct sides of the microphone.

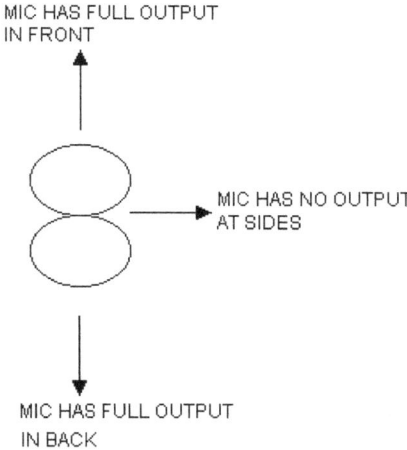

Bi-Directional Pickup Pattern

The hyper-cardioid microphone pattern has its maximum pickup from the front of the microphone and less from the sides - similar to the cardioid pattern. Its minimum pickup is not directly in the back but at the side of the diaphragm's rear. The minimum pickup is about 120 degrees from the front of the microphone. The hyper-cardioid pickup pattern is designated with a heart that has a small tail in the back.

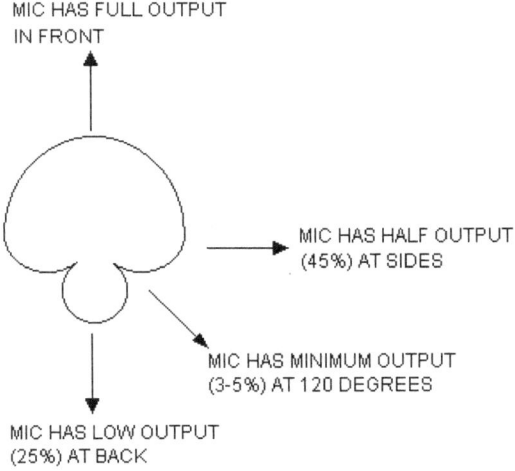

Hyper-Cardioid Pattern

There are some microphones that have more than one pickup pattern. These microphones actually have more than one diaphragm in the case. There is a switch on the microphone that allows the user to change the microphone pattern. The switch activates one diaphragm, the other or a blend of the two diaphragms to get the different patterns.

Pointing a Directional Microphone

The important part of pointing a directional pattern is to point the rejection of the microphone at what you don't want to pick up.

Microphone Pattern Demonstration

Your instructor will put a multiple pattern microphone out in the studio and have different students talk into the front, sides and back of the microphone when it is set to different patterns. This will allow you to hear the effect of different microphone patterns.

PAS DEMOS Students having access to the PAS Demos, Listen to Demo 3, Microphone Patterns.

Close and Distant Micing Techniques

There are two distinct types of microphone placement techniques. The close micing technique involves putting the microphone less than a foot away from the sound source. Close micing gives maximum isolation. In the studio, close micing is used much more often than distant techniques.

Distant micing techniques involve putting the microphone 3 feet or more away from the sound source. Distant micing techniques are used to pick up a lot of musicians with just a few microphones and are used when the room that the musicians are playing in sounds "good."

Distant micing is the most natural micing. Most instruments are made to play to people more than three feet away. When a microphone is placed very close, often certain notes will sound different than other notes (brighter or louder). Mechanical sounds like the "squeaks" of a foot pedal or the fingers running over the neck of the guitar are more pronounced when the microphone is placed close to the instrument. Even with these disadvantages, close micing is used more often because of its big advantage: isolation!

Close Micing	*Distant Micing*
• Mic Less than 1 foot away	• Mic 3 feet away or more
• Best Isolation	• Poor Isolation
• More Presence in Sound	• Less Presence, Fuller Sound
• Less Even Tone	• More Even Tone
• More Hyped Sound	• More Natural Sound
• More Mechanical sounds	• Less Mechanical Sounds
• Less Room Sound	• More Ambience
• Isolates one instrument from group	• Picks up sections of instruments

Comparing Close vs. Distant Micing

Close and Distant Micing Video

Your instructor will play a demonstration video of a guitar and piano with close micing and distant micing. As you are listening to the close mic, notice how other sounds in the room are diminished (the student who talks to the musician), how there are accented sounds of the fingers moving over the guitar neck and how the middle range of the piano is accented. When listening to the distant microphone, notice how there is distinct room reverberation, how the notes have a more even tone and how the student talking to the musician sounds louder.

CHAPTER 4 – MORE ABOUT MICROPHONES

Cycles per Second

Remember that sound is a wave that, if it were seen, would look like a wave in a pond. A *Cycle* is one complete crest and fall of the air in the wave - one complete movement of the air particles. One cycle is made by one complete vibration of the vibrating source.

Sound Wave

To make sound, the vibrating source must vibrate between 20 times a second and 20,000 times a second. This causes the wave to be between 20 cycles per second and 20,000 cycles per second. The rate of cycles per second is called the *frequency*. The unit of frequency is the *Hertz*. One hertz is one cycle per second. Thus the frequency range of sound is 20 hertz top 20,000 hertz.

Frequency Ranges

The different frequencies of sound are divided into ranges.

The **Bass** frequencies are the frequencies from 20 hertz to about 350 hertz. The **Midrange** frequencies are from about 350 hertz to 4,000 hertz. The **High** or **Treble** frequencies are from about 4,000 hertz to 20,000 hertz.

PAS DEMO

Students having access to PAS Demos, Listen to PAS Demo 3, Frequency Ranges.

Wavelength

Sound travels at about 1130 feet per second. Sound of different frequencies would have a different number of cycles fit into 1130 feet. In a 100 hertz sound wave, there would be 100 cycles in 1130 feet. In a 10,000 hertz sound wave, there would be 10,000 cycles in 1130 feet.

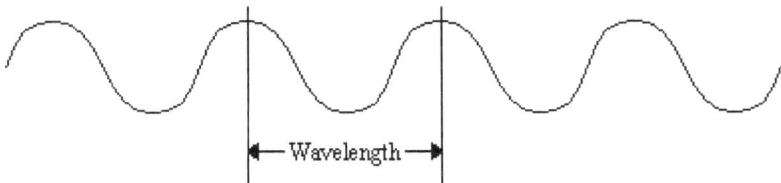

The wavelength is the length, in feet or inches, from the crest of one wave to the crest of the next wave. The wavelength of a 100 hertz sound wave would be about 11 feet 4 inches (11.3 feet). The wavelength of 1000 hertz would be about 1 foot, 1 inch (1.13 feet). The wavelength of a 10,000 hertz sound wave would be a little over 1 1/4 inches (.113 feet)

The bass frequencies have longer wavelengths, in the range of about 10 feet. The midrange frequencies have medium wavelengths, in the range of about 1 foot. The high frequencies have very short wavelengths, in the range of 1 - 2 inches.

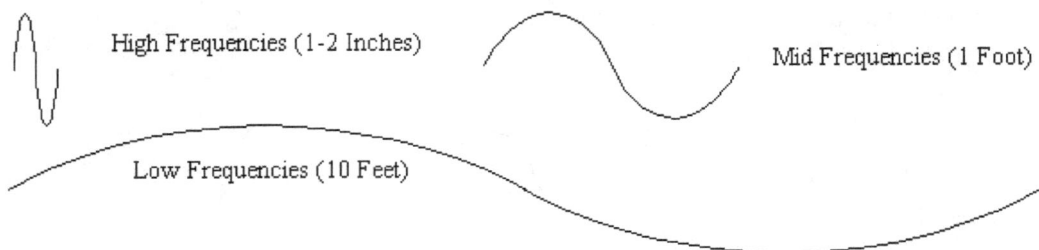

The wavelength difference of sound causes the sound to act differently and to be picked up differently by microphones. The differences are more noticeable at the longer wavelengths of the bass frequencies. The long wavelengths of bass frequencies require thick walls to stop the sound wave. If you hear music playing in the next room, you will hear the bass frequencies rather than the high frequencies. Lower frequencies, with their longer wavelengths will travel further than the high frequencies. The "crack" of lightning has a lot of high frequencies; but at a distance it sounds like bass-frequency thunder.

The Standing Wave

When the distance between two reflective walls is close to the length of one or two cycles of a sound wave, a standing wave can occur. The standing wave is a wave that sounds like it is standing still. What happens is that the reflection of the sound wave causes the crests of the sound wave to be in the same location of the room as the original sound wave.

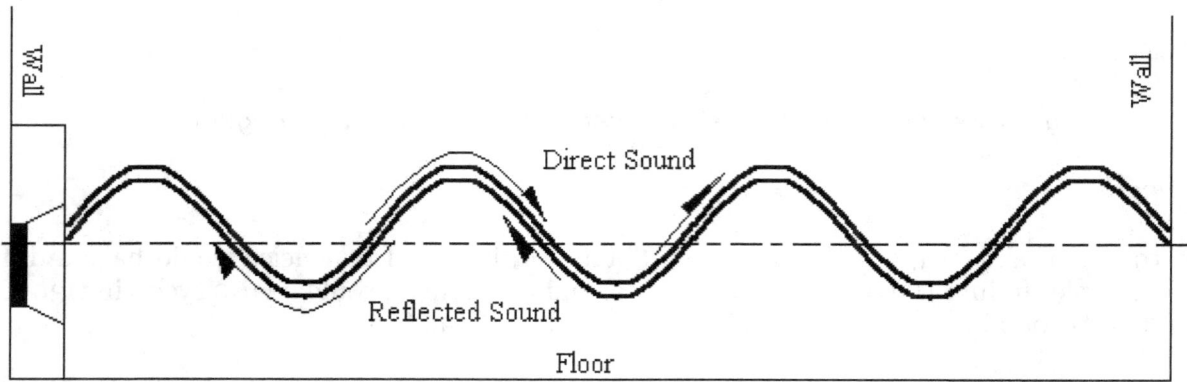

Standing Wave

Standing waves only occur at very low-frequency sound waves with very long wavelengths. As you walk along a sound wave, you will hear very loud bass then very soft bass and then very loud bass again. This sounds like the crests and valleys of the sound wave cycles.

Chapter 4 – More About Microphones

Standing Wave Demonstration

Your instructor will put a low-frequency wave in the studio. The frequency of the wave will be 40 hertz, approximately the frequency that would be put out by the lowest string of an electric bass. As you walk away from the sound source, you will hear the bass get loud and then soft. The instructor will measure the distance between two loud portions to give you the idea of wavelength.

Harmonics & Fundamental Frequencies

When sound is put out by instruments, the pitch of the instrument is "tuned" to a certain frequency. This "tuned" frequency is called the fundamental frequency. The instrument will put out energy at whole-number multiples of the fundamental frequency. These multiples are called harmonic frequencies. The harmonic frequencies is what makes one instrument sound different than another instrument playing the same fundamental frequency.

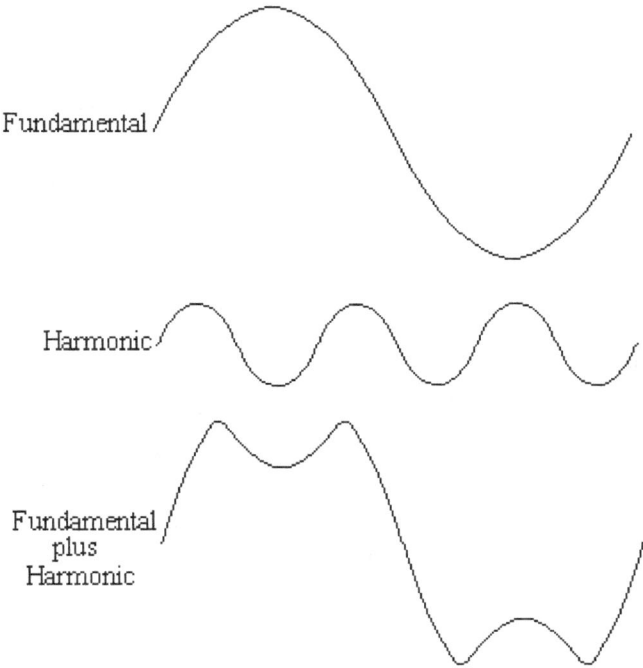

Fundamental & Harmonic Frequency Video

Your instructor will play a video for you that will have several instruments play the same pitch. The pitch on each of these instruments will have a fundamental frequency of 110 hertz. The screen will display the waveform that is put out by the instrument as it is sounding.

Microphone Placement for Accenting Harmonic Frequencies

The harmonic frequencies give the clarity and recognition of the musical instrument. The harmonic frequencies should be accented in recording. The video will show how an acoustic guitar has increased harmonic frequency content when it is miced over the end of the strings, at the bridge. Any stringed instrument will have increased harmonics at the end of the vibrating string.

Proximity Effect

Proximity Effect happens in directional microphones, such as microphones with a cardioid pickup pattern. The proximity effect is an increase in bass frequency output when the microphone is placed a foot or less away from the instrument.

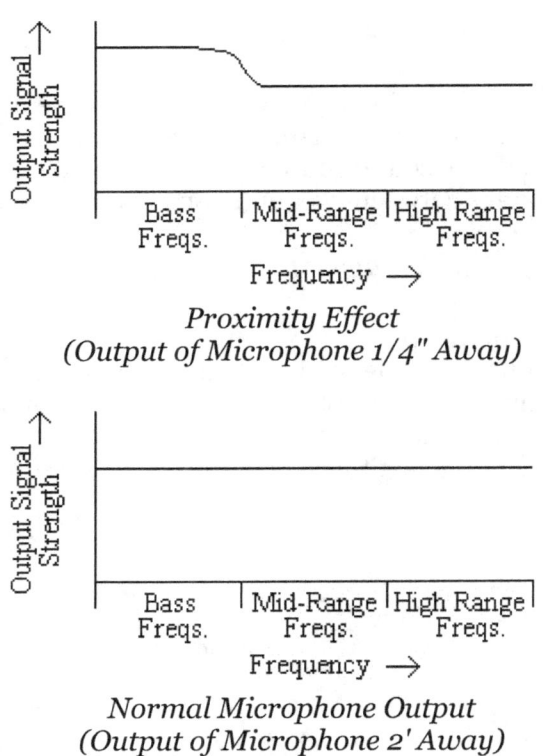

Proximity Effect
(Output of Microphone 1/4" Away)

When a microphone is placed 1/4 inch away from the instrument with a cardioid microphone, the very low-bass energy is picked up about four times louder than it should be because of proximity effect. The rest of the bass frequencies are accented to about three times normal energy. When a microphone is placed 1/4 inch away from an instrument, the microphone sounds like someone turned the bass control on the stereo all the way up.

Normal Microphone Output
(Output of Microphone 2' Away)

If an instrument sounds good, the proximity effect of the microphone can cause the recording to sound muddy. Microphone manufacturers use three different methods to correct the proximity effect:

1. Some manufacturers make microphones that permanently have reduced bass pickup to compensate for the proximity effect. Usually the bass energy is reduced to about 40% of what it should be. These microphone sound "correct" when they are placed 3 to 6 inches away from the sound source. When they are placed closer, they sound a little to bassy. When they are placed further away, they sound thin.

Often microphones that are designed for live sound reinforcement have this characteristic. Examples of these microphones are the Shure SM-57 & 58 as well as the Beta 57 and 58 models.

Some studio microphone also have the permanent bass roll-off and are designed for close-micing only. The Neumann KM series microphones (KM 85 and KM 86) are good examples of this.

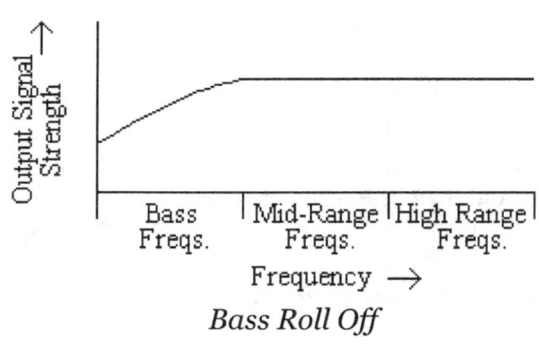

Bass Roll Off

2. Many manufactures have switchable bass roll-off. When using this type of microphone the bass roll-off is turned on when you place the microphone closer than about 8 inches and turned off when the microphone is placed further from the microphone.

The vast majority of directional microphones designed for studio use have switchable bass roll-off. Many manufactures have two position bass roll-offs where the extreme roll-off sounds best when the microphone is placed less than 3 inches away, a mild bass roll-off position that sounds best when the microphone is placed 3-8 inches away, and the no roll-off position for micing more than 8 inches away.

3. Cardioid microphones are made directional by putting ports (holes) in the case behind the diaphragm. To work correctly, the ports sound be different distances for different frequencies. The bass frequencies, with their longer wavelengths, should have ports further back on the case.

In the 1950s, ElectroVoice invented a microphone case that helps reduce the proximity effect. The invention is called the Variable-D (Variable-Distance) port system. These microphones have ports which extend 4 to 6 inches down the microphone's case. There is a wedge-shaped piece of foam put behind the ports so that the further ports only allow the bass frequency sound waves to pass through the ports. This puts ports on the microphone case further back from the diaphragm for bass frequencies.

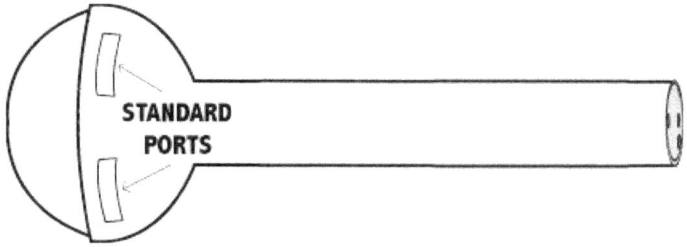

Regular (Single-D) Cardioid Microphone Case

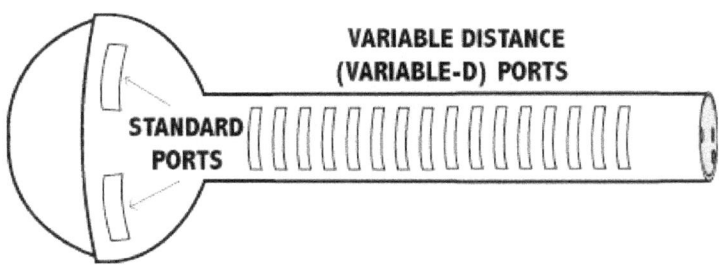

Variable-D Microphone Case

The Variable-D microphone has much less proximity effect and only increases bass pickup by less than two times normal when the microphone is 1/4 inch away. The Variable-D microphone will therefore pick up the correct amount of bass anywhere from 2 inches to several feet away from the instrument. The famous RE series studio microphones (RE-10, RE-15, RE-16, RE-20 and RE-27) are examples of microphones that used Variable-D ports. The RE-20 and RE-27 are still available and widely used. Some other manufacturers also paid ElectroVoice for the right to use Variable-D ports before their patent ran out.

Some models of microphones do not have any method of compensating for the proximity effect; these models do not use a bass roll-off or Variable D port system. When using directional microphones that don't have bass roll-off or Variable-D ports, you should reduce the bass at the console for close microphones.

Proximity Effect Demonstration

PAS Audio Demonstration – Students having access to PAS Demos, Listen to Demo 4, Proximity Effect.

Your instructor will play a recorded demo for the class and adjust the board's equalizers so that you can hear what proximity effect does to the sound.

Phase Cancellation

The term phase refers to two similar waveforms slightly out of time with each other. Two waves are said to be in-phase when the crests of the two waves occur at the same time. Signals that are in-phase will add to make a signal with twice the strength. Two signals are said to be out-of-phase when one wave's crest occurs at the same time as another wave's valley. Two waves that are out-of-phase will cancel each other out when they are mixed together.

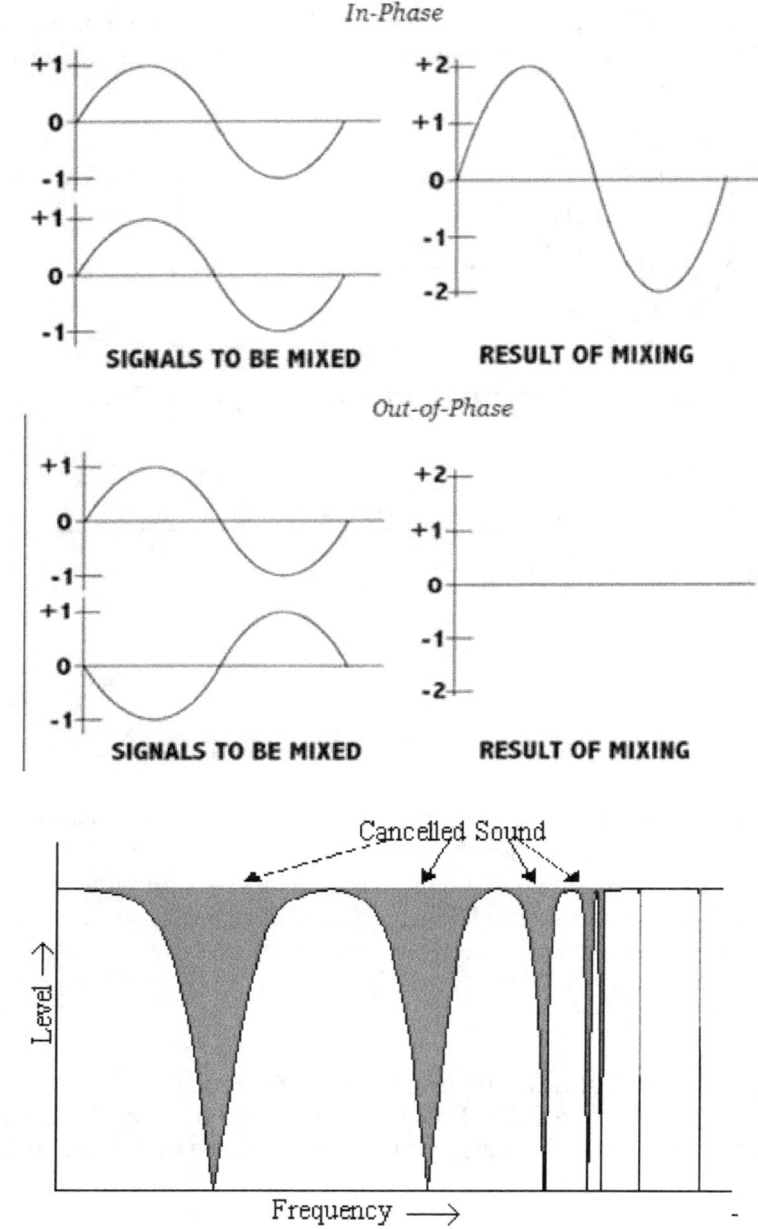

Phase Cancellation from Multiple Microphone Interference

Phase is similar to two people pushing on a door. If two people push on the door the same way (in-phase), the door opens twice as fast. If one person pushes on the door as the other person pulls (out-of-phase), the door will not move.

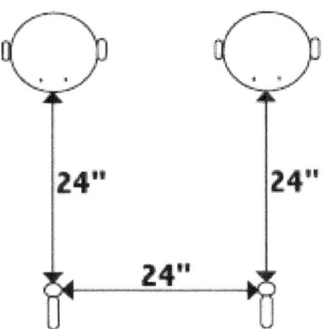

When placing microphones, signals can get out-of-phase when two microphones pick up the same instrument from different distances. When this happens, some of the instrument's frequencies cancel and others add. This makes the sound of the instrument very thin and unnatural. The sound is said to have phase cancellation due to multiple microphone interference. Phase cancellation is most noticeable when two microphones are placed about 6 inches from each other, both picking up the same instrument.

Three To One Rule

When microphones are placed far enough apart, the leakage into the second microphone is not strong enough to cause phase cancellation. Part of picking up a good sound is to prevent phase cancellation. One way to prevent phase cancellation is to observe a simple microphone placement rule called the Three to One Rule.

THE THREE TO ONE RULE STATES THAT A SECOND MICROPHONE (INTENDED FOR ANOTHER INSTRUMENT) MUST BE PLACED THREE TIMES FURTHER AWAY THAN THE MICROPHONE IS FROM THE SOUND SOURCE

Let's say that you are micing two singers. If a microphone is 8 inches away from one of the singers, than the microphone for the second singer must be at least 24 inches away from the first microphone. Often it is a good idea to keep the second microphone even further away so that the singers can move slightly without causing phase cancellation.

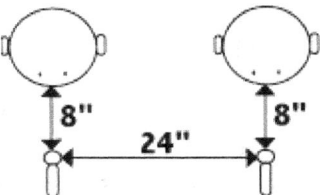

Three-to-One Rule

Phase Cancellation Demonstrations

Your instructor will play a demonstration for you that shows you the sound of phase cancellation. On the recording, you will hear part of a tune, as it was intended. The second playing of the tune will have one side slightly delayed. As long as the sound comes from the two speakers you will not notice a big change in the tone. Half way through the second playing the sound will become mono (left and right channels mixed together). When this happens, you will notice a big change in tone where much of the sound disappears and the tune sounds thin and unnatural.

Your instructor will place two microphones on a singer, about 6 inches away from each other. When the two microphones are mixed together at the same volume, the voice will have phase cancellation. An assistant will move the second microphone much further away from the first and the voice will again have normal tone.

Direct Pickup Recording

Electric instruments, such as the electric bass and the electric guitar, do not put out sound. You hear these instruments because they are connected to an instrument amplifier. The electric instruments put out an audio signal that is amplified and sent to a speaker in the instrument amplifier.

Many times it is undesirable to have loud electric instruments playing in the studio, leaking into other microphones. This is especially true of recording in a home studio that does not have isolation rooms.

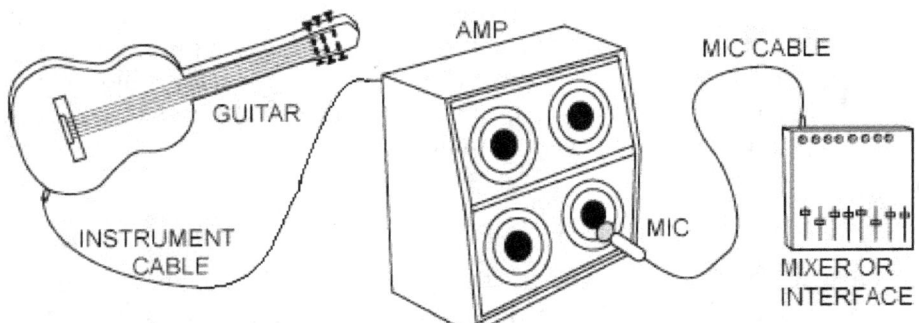

An electric instrument puts out a strong audio signal. The strength of the signal is called line level. A microphone puts out a much weaker audio signal, called a microphone level signal. For a relatively small amount of money (about $50) you can purchase a direct box that will convert the line level signal (out of an electric instrument) to a microphone level signal. This eliminates the loud instrument amplifier in the studio. The signal still reaches the console and can be recorded.

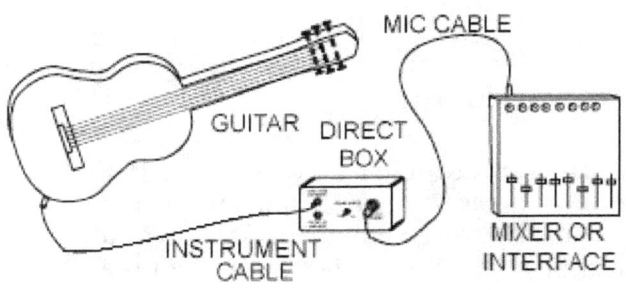

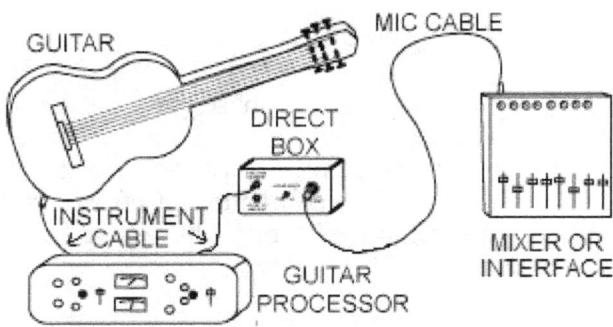

Bass guitars are most-often recorded direct (using a direct box) in a studio. Electric guitars, however, usually do not sound right when they are recorded like this. This is because much of the sound of an electric guitar is obtained by the instrument amplifier. Several manufacturers make accessories called "guitar processors" which can make a direct guitar sound like it was placed through an amplifier. These guitar processors cost between about $150 and $1000 and can be used for recording the guitar (or bass) direct or can be used to make an instrument amplifier sound different. Also available are Amp-simulating plugins.

CHAPTER 5 – USING A BASIC RECORDING SYSTEM

In the days of old (last decade), basic recording was done using microphones sent to a console, to a recorder, back to the console, to the speakers and the headphones.

In modern recording, the recording and mixing of signals are both done within a computer program accessed with an audio interface, a mouse and a keyboard. This setup has replaced most functions originally taken care of by a console and a recorder.

Regardless of the fact that consoles are not in general use in basic recordings systems, a basic understanding of their function is necessary because that is the basis for interface and computer operation today.

What is a Console?

The console was at the heart of the control room operations. A console is essentially a router and a mixer. Similar to a telephone switchboard, it can take in many signals and then send them out to different places. The function of this is called *routing*.

The second main function of a console is mixing. The mixing function of the console is taking two or more incoming signals and combining them into one signal. Originally, the recorded sound may have come in from 16 tracks of the recorder. The 16-track recorder will put out 16 different signals. These signals have to be mixed into a final master recording, which we used to send to a 2-track recorder. Once the final master recording is done, there was two signals recorded, a left signal and a right signal.

Basic Recording Setup, Last Decade and Before

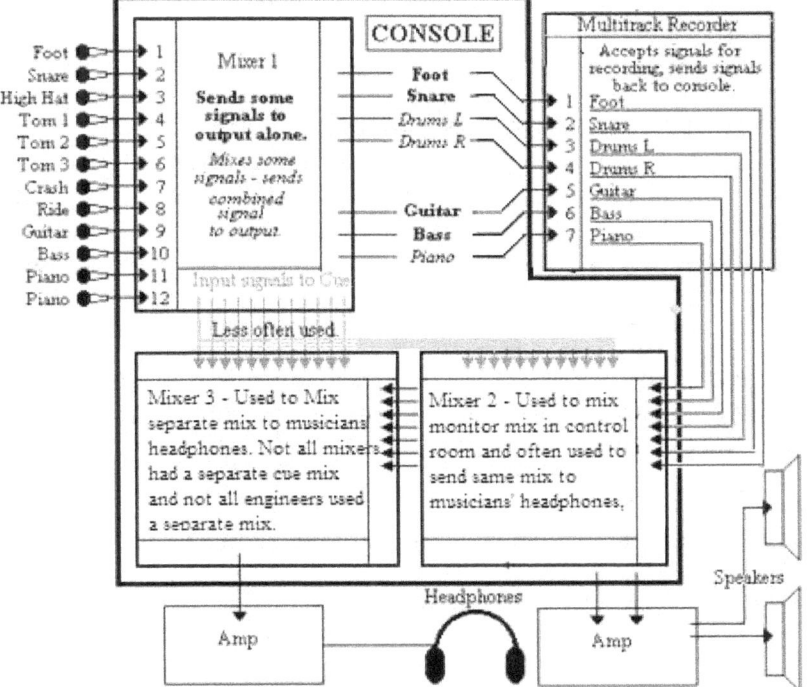

The console actually had 3 or more mixers, all in one housing. During a recording session, three mixers were to be used:

Mixer #1:

This mixer will be used to combine microphone signals together so that more than one microphone can be recorded onto one (or more) tracks. Some microphone signals can go directly to one channel, but others have to be mixed together to be recorded.

For a 16-track recording session, the drums would probably be recorded onto five tracks. The three microphones for the kick drum, the snare drum and the high hat cymbals would be recorded onto three individual tracks by routing each of them directly to a track. The remaining drums and cymbals would take 5 microphones, and would be recorded onto two tracks that are labeled *Drums Left* and *Drums Right*. So the engineer would use one of the mixing sections of the console to mix 5 microphones to two tracks.

Mixer #2:

We were recording onto a 16-track or 24-track recorder. Most of us don't have 16 or more ears - we have only 2. We hear the audio by listening to two speakers, one for each ear. We still have to hear all of the instruments that are being recorded. The second mixing function is to mix together the audio signals coming from the 16-track tape recorder and send them to the amplifiers that drive the speakers. This mixing section of the console is called the *monitor mixer*.

Mixer #3:

During the recording sessions, we have musicians that have to hear the other musicians, even if they are in other isolation rooms. After the initial recording, musicians will record other parts onto other tracks in the DAW program or tape (a process called *overdubbing*). When overdubbing, the musicians must hear the previously recorded tracks plus their own performance in headphones. The engineer must use another mixing section to mix the signals that the musicians hear. The signal that the musicians listen to is called the *Cue* signal (because it gives the musicians their cue to play). The monitor mixing section that the engineer uses to mix this signal is called the *Cue Mixer*.

Operating a console during a recording session is something like an air traffic controller that would have to operate three planes at the same time from the ground. They have controls for each plane in front of them. Before turning the wheel for a right turn, the controller would have to make sure they were reaching for the controls for the correct plane. Almost all confusion of the student at a recording console comes from keeping the controls straight - Which controls are for which function?

Modern Recording

Modern recording is done with the use of a DAW program. DAW is short for Digital Audio Workstation, which is a unit containing the channel controls of a console and the recording tracks, all housed within the same unit.

There was a short time where DAW units were available, but computer programs replaced them pretty quickly and the name is now almost exclusively applicable to such a program.

With most DAW programs having 48 tracks or more available, the modern recording engineer almost never has to mix inputs to tracks, The author asked a recording engineer in a professional studio who has done recording almost continuously in the last decades and when asked the last time he mixed to track, he said almost in the last millennium.

Chapter 5 – Using a Basic Recording System

Input Levels

During recording, the main channels of the recording system are used for picking up, adjusting levels and routing the signals from the microphone inputs. The first thing that the engineer must adjust is the input level to the system. Too high of a signal will cause input overload and make the signal distorted (garbled). Too low of a signal will cause noise (or more noise than would be had had the signal input been set properly). The engineer has a microphone gain control that he/she uses to adjust the signal as high as possible without overloading the input.

Some recording studios use separate input amplifiers which are sent to the interface connecting to the computer DAW. The more inputs a studio has, the more likely the input amplifier and the interface will be separate devices.

Most modern preamplifiers, whether within an interface or within a separate unit, have overload or *clip* lights that will begin to flash when the signal is approaching overload. The light usually begins to flash when the signal strength is about 75% of the signal strength that will cause overload.

The engineer will have the musician play at the highest volume he/she will be playing and adjust the microphone gain control so to the highest position that can be obtained without the light flashing. The level indicator within the DAW track should read -6 dB.

Once the musicians start running though the tune several times, they often increase the level they are playing at by about 25%. During the recording, it is normal for the overload light to flash occasionally. The light, however, should not be continually on, or come on for each note sounded.

When the system is routed and set for recording, the signal strength of the channel path will read near -6 dB for digital. If there is an input fader in use, the fader will normally run at about 75% up. This position is normally marked proper level.

Having to set the channel volume control too high or too low to get the proper level in the recording system is an indicator that the microphone gain control is set incorrectly.

When dealing with a separate console and console meters, the diagram on the next page shows both the correct and incorrect settings.

Input Overload/Noise Demonstration

PAS DEMOS Students having access to the PAS Demos can listen to PAS Demo 5, Input Overload.

Your instructor will set the levels correctly for recording a snare drum, according to normal procedures. With "0" coming out of the console, the channel fader will be close to 75% up.

After correctly setting the microphone gain, the instructor will show you the effect of incorrect settings. First, the instructor will purposely set the mic gain too high. The level will come up and he will adjust down the channel fader to 50% or below so that "0" level still comes out of the console. The levels will look right out of the console but the overload light will stay lit and the sound will be close to someone hitting trashcans in the alley. The instructor will then purposely set the microphone gain way to low. The instructor will have to bring the fader all of the way up and perhaps adjust other console controls way up to get "0" level out of the console. Although the snare will sound pretty normal, there will be a distinct hissing sound present in the channel.

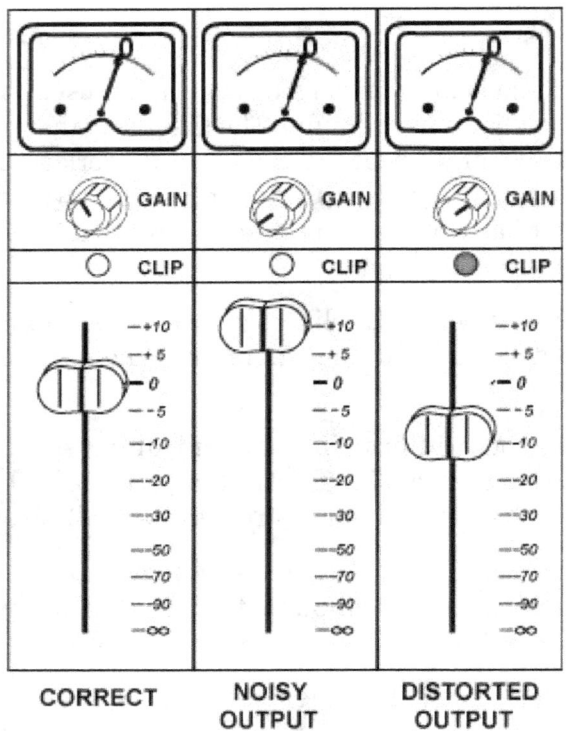

CORRECT — NOISY OUTPUT — DISTORTED OUTPUT

The Digital Recorder and Its Levels

Much of the professional recording done today is done digitally within DAW programs.

In the digital system, the audio signal coming in is measured several thousand times a second (usually 44,100 or 48,000 times a second). Each time it is measured, the audio signal level is assigned a number to define the level. Pulses are then recorded onto the hard disc, which represent the numbers. Since the audio signal levels are converted to numbers, this is referred to as a digital recording.

Digital systems have the distinct advantage of having no tape noise and of copies sounding like the original.

Digital tracks cannot tolerate any amount of overload. When the signal coming into the interface is too high, the machine runs out of numbers to define the audio level. The result is immediate and severe distortion. The recorders are sensitive to the peak energy of the highest crest of the waveform and the digital recorders meters read this peak energy.

Console meters, when routed into used DAW program interfaces rather than using a rack amp unit, often respond to the average level of the waveform, which determines how loud it sounds.

There is can be quite a difference between the peak energy of the waveform and the average level of the waveform. Instruments like bass guitars have low level of peak energy compared to the average level. Instruments that are percussive, like drums and cymbals, have a great amount of peak energy.

Chapter 5 – Using a Basic Recording System

After you adjust the console for the correct levels at the console input and output, you will have to adjust the levels so that the digital recording machine's meters read about "-6." The lower level of "-6" is used to provide room for an unexpected peak to occur and still not go over the "0" point. If any peak goes over "0" it will be severely distorted. To get the digital recorder's meters to operate, you must arm tracks that you are going to record. When the tracks are armed, there will be a flashing red light under the track meter and you will see the input-signal peak-level on the track's meters.

You will find that the levels of percussive instruments, such as drums and cymbals, will have to be turned down. The console meters may read as low as "-10" when the levels are adjusted to the correct level for the tape machine. To reduce the level for these instruments, reduce the main channel fader below the 75% area. Analog console meters may read as high as "+3" for melody instruments such as bass guitar, to get the correct levels for the digital recorder. Change the main-channel fader position to slightly above the 75% level as necessary to get the proper recorder levels.

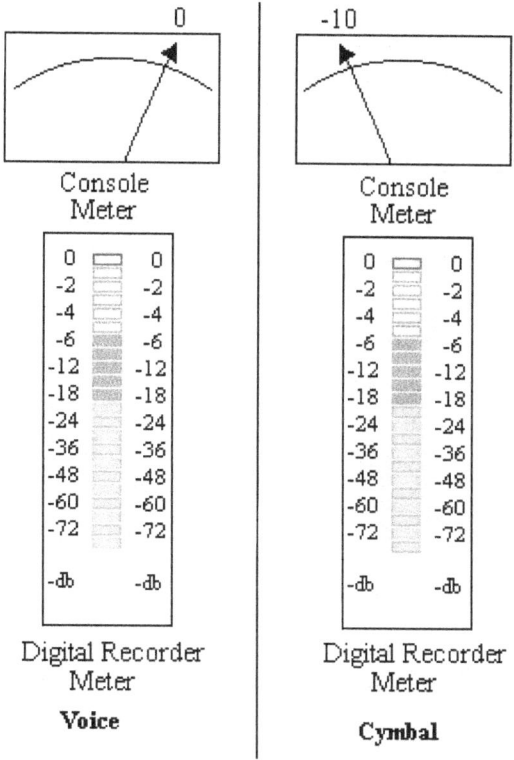

Correct Digital Recorder Levels

Peak and Average Level Demonstration

The instructor will open a DAW program meter plugin which shows average and peak level. This plugin should be added to a kick drum and bass guitar. The student should note the level differences and comparative volume of the kick vs. bass guitar.

Monitor & Cue Mixing

Keeping It Straight:

In traditional recording, a console was used to mix signals to tracks creating a situation where the numbers didn't match up.

When such a system is used, care must be taken to apply the signal processing and other aspects to the correct channels, whether input or return. For ease of keeping it separated:

WHEN MIXING MONITORS AND CUES, THE ENGINEER USES THE TRACK NUMBERS.

Routing Microphones to Tracks

Input Selection:

Some interfaces allow up to three different possible signals to be sent to track: Microphone, Line, or DI (Instrument level).

To prevent user error, most interfaces have separate inputs for mic and line. This prevents phantom power accidentally being sent back to a device with a line output, this damaging an output. Some interfaces can use the same input, but only a microphone's connection can be sent phantom power.

In the More on Microphones chapter, it was explained that guitars can be sent directly to an interface using a DI box.

Because the input is handled differently for line and instrument, a button is usually activated when using the input as a DI-in.

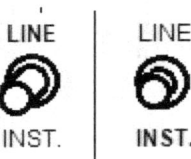

Setting Line vs. Instrument Input

Input Selection Switches

On some larger consoles, line inputs were used as tape returns and the engineer would have to set recording or mixdown status to adjust which inputs went to the main channels and which inputs went to the monitor.

If the engineer needed a line input in recording status, there was input selector to allow that.

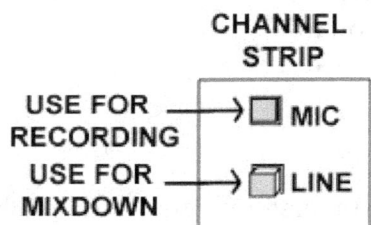

Direct Routing

Direct routing means that one microphone signal is sent to one track. On any console in use, the microphones are sent to the same interface number and therefore track number as the microphone input number. In other words, activating the direct routing switch on strip 5 will cause the microphone signal for mic input 5 to be sent to DAW track 5.

Busses

As noted before, it is almost unheard of for signal to be mixed to track in modern recording.

If it is done, it is done with a buss. All mixing in consoles (and in DAW programs) use a buss. A buss is a wire that can accept different signals and combine them, or a computer function which mimics that.

Common buss types are recording busses (for recording), auxiliary busses (for effects), cue busses (for musicians' headphones) and monitor busses (for the control room).

Anything labeled as "Output Number ___" in a DAW program is actually being sent to a buss which is then sent to that physical output. Even the main output is a buss sent to the main outputs.

The master fader in a DAW program is effectively a main-buss control channel.

Chapter 5 – Using a Basic Recording System

All busses and outputs can have a master fader activated allowing the engineer to control it separately from the channels.

This is almost never done.

The busses in a console will have master controls that adjusts the level of the overall mix of signals sent to the tracks. The name of these control vary from console to console. Some of the names used are: Buss Master, Buss Gain, Buss Trim, Group Master, Group Gain or Group Trim.

Setting Levels While Using Busses

Procedures on getting levels change slightly when you are routing several microphones to one track. The input levels, set with the overload light, is the same. To set console levels, you set the microphone levels for each of the inputs one by one. When doing this the buss master level should be all the way up, the channel fader should be at about 75% to get "0" level on the console meters.

After you check all of the individual levels, you have all of the musicians perform. Adjust the individual channel faders so that the mix of the musicians is correct (usually equally loud). If the console meter reads too high, reduce the buss master to get the correct level at the console. Now you set the level for the track send. Reduce the buss master level, if necessary, to get the digital recorder's track level at the "-6" level.

Using Pan Pots When Routing Microphones

Sometimes you want to record several microphones to two tracks to make a left and right image. An example of this is the Drums where 5 microphones may be routed to two tracks labeled Drums Left and Drums Right.

To let you do this there will be a small pan pot control that is associated with the buss buttons. Usually, there will be a pan switch that activates the pan pot used for routing. This routing panpot will pan between odd and even busses that feed odd and even tracks. Odd busses and tracks are considered left, while Even busses and tracks are considered right. To use the pan pot you must send the microphone signal to both an odd and an even buss. In our example for drums, we had the following microphone numbers and track numbers:

Microphones:

5. Small Mounted Tom
6. Larger Mounted Tom
7. Large Floor Tom
8. Overhead Cymbals (Left)
9. Overhead Cymbals (Right)

Tracks:

5. Toms & Cymbals Left
6. Toms & Cymbals Right

To route these microphone to the tracks we would do the following:

- *Strip#5: Push buss buttons 5 & 6, Push Pan Button, Move Pan Pot Clockwise*
- *Strip#6: Push buss buttons 5 & 6, Push Pan Button, Move Pan Pot to Center*
- Strip#7: Push buss buttons 5 & 6, Push Pan Button, Move Pan Pot Counter-Clockwise
- Strip#8: Push buss buttons 5 & 6, Push Pan Button, Move Pan Pot Counter-Clockwise
- Strip#9: Push buss buttons 5 & 6, Push Pan Button, Move Pan Pot Clockwise

Set Up Sequence

There is a sequence of actions to routing the console and getting levels for the session as follows:

1. Zero the equipment: There are many controls, switches and buttons on a console or interface. You don't want the controls to be in the wrong position because they are still set up for the last session. Turn all controls down, off or in the "normal" center position.

2. Put monitor controls in a "rough position: Turn the monitor level control up to about 50% for every track you are recording and any previously-recorded tracks. Adjust the pans to taste. This allows you to hear any track as soon as you route it.

3. Rough set the cue controls: Do the same thing for the cue mix controls so that the musicians will hear themselves as soon as the microphone is routed and they begin to play.

4. Get the recorder ready: Put the tracks to the correct starting point and arm the tracks you are going to record. This gives you meters to set levels with and allows the monitors and cues to get signals.

5. Set levels: One at a time, have the musician play. Make sure the console is on "mic" for that input signal and the 48-volt switch is activated., if needed. Adjust the mic gain control so there is no overload, adjust the channel fader (or buss master) for the correct recorder levels.

6. Rehearse: Have the band go play the tune they are going to record. Make fine adjustments to mic gain, channel/buss levels, and monitor levels and cue mix levels as needed.

7. Record!

CHAPTER 6 – MIXDOWN

What Is Mixdown?

The mixdown is the process of mixing all of the recorded tracks down to a stereo master recording.

To get this started, the main channels reduced down to their lowest levels. All non-production settings are removed allowing a fresh start.

Effect channels are added to the DAW program and are set to Solo Defeat so that they can be heard during Solo monitoring.

The main output of the DAW is sent to the monitoring system being used for mixing.

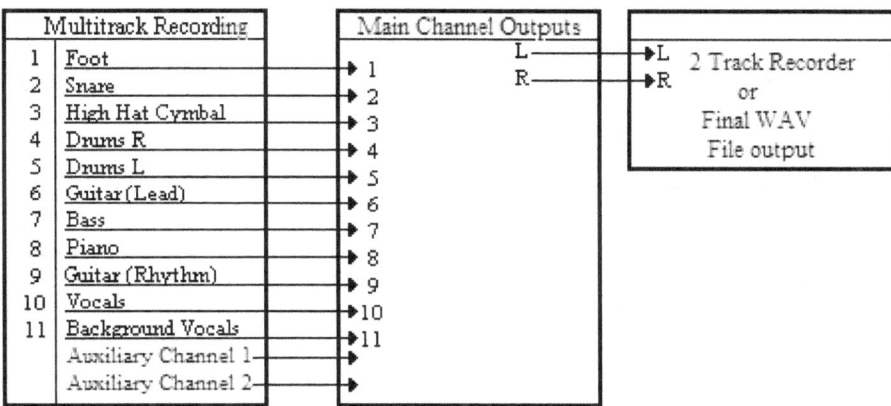

When mixing, several important factors have to be remembered and taken into account.

master echo return control to get the same amount of reverberation.

Hearing Limitations

Ears are not the best things to listen with. Unfortunately, they are the only things we have. There are conditions under which you will not accurately hear the sound as it is being recorded. These "hearing limitations" really can affect the mixdown. This means that the engineer has to understand the problem and get ways around it. Overall, there are five hearing limitations.

If you look at the list, the majority of hearing limitations are overcome by accenting harmonics. This is usually done with equalization, explained later in the chapter.

Pitch Change

      A Condition of the hearing where at loud volumes bass sounds flat. If a singer is singing off key flat, turn down the headphone or turn down the bass.

Hearing Distortion

> The nerve endings in the inner ear can distort if overworked. Do not mix at too high of a level for too long and if you are hearing any sort of buzzing, stop mixing and go to a quiet environment and let your ears rest.

Bass Directionality

> The human ear cannot hear direction below 200 Hz because the waveforms are too long and hit both ears at the same strength. To allow instruments with low end (e.g. guitars) to have direction when panning, accent harmonic frequencies.

Masking

> As explained in More or Microphones, instruments that are too similar can cover each other up. Overcome this by accenting different harmonics using microphone placement or equalization.

Fletcher-Munson Effect

The biggest hearing limitation was researched by Fletcher and Munson. This limitation is called the Fletcher-Munson Effect.

Fletcher and Munson discovered that when music is played low, it is very hard to hear the extremely low bass frequencies and somewhat harder to hear the extreme treble frequencies. The lowest bass frequencies are about 64 times harder to hear and the extreme treble frequencies are about ten times harder to hear, at playback levels near the soft conversation levels. The ear hears the frequencies most correctly when the music is played fairly loudly.

THE FLETCHER MUNSON HEARING LIMITATION IS THAT IT BECOMES VERY HARD TO HEAR THE LOW BASS AND SOMEWHAT HARD TO HEAR THE HIGH TREBLE FREQUENCIES WHEN THE MUSIC IS PLAYED BACK AT A LOW VOLUME.

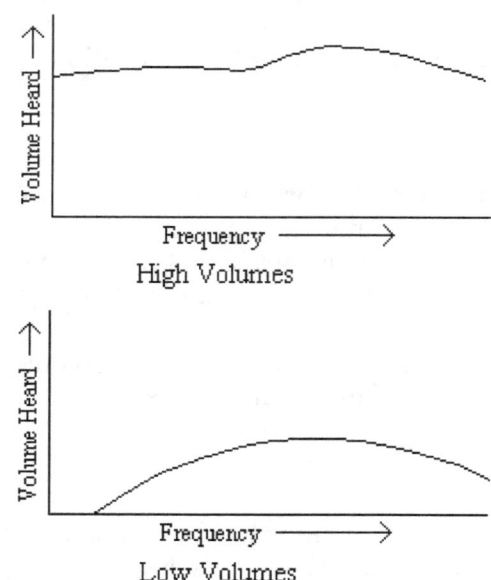

You can't get around this by simply playing the music loudly as you are mixing. Your objective is to get the person buying the record and hearing the record to be pleased with the sound. People listen to music at all different volumes. The same person will listen to music at different volumes at different times, depending on circumstance. The object is to get the final product to sound the best under all listening circumstances.

The first thing that the engineer needs to do is to listen to the mix at *different volumes*. As soon as the mix sounds good to you, listen to it at a lower and at a higher volume. This will allow you to make adjustments for the best sound under all circumstances.

If, during the recording process, the harmonic frequencies were accented by proper microphone placement, the Fletcher Munson Effect will be easier to get around. The Fletcher Munson Effect does not affect the harmonics of a bass line as much as it affects the fundamental frequencies.

Fletcher Munson Demonstration

The instructor will play music loudly and softly for you. He will direct you attention to how easy it is to hear the bass, voice and cymbals when the music is played at different volumes.

The Mixdown Perspective

Left, Right and Center

During mixdown you are painting a picture of the musicians performing for you. You are trying to get the perspective of performance.

One of the perspectives is the left, right and center position of the musicians' instruments. The major control which allows the engineer to effect the perception of left to right is the pan pot. The main channel pan pots are used to place the instruments sound like they are coming from left, right and center stage.

Distance

The other perspective that you have to deal with during mixdown is the perspective of the distance that the instrument is away from the listener.

When an instrument plays in a room, it puts out a sound wave that travels to the listener. This is the "direct" sound - the sound directly from the instrument to the listener. The sound also bounces off of walls, ceilings and floors to make reverberation. When a person hears an instrument in a room, he/she hears both the direct sound and the reverberation. The blend of how much direct sound there is to how much reverberation there is determines the perspective of distance.

As a person walks away from an instrument, the direct sound diminishes. The reverberation is approximately the same any place in the room (for most rooms). So the result is that close instruments have a higher percentage of direct sound and instruments further away have a higher percentage of reverberation.

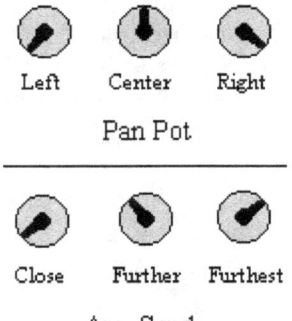

Controls for Mixing Perspectives

The Reverberation Controls

Remember that when we recorded, we often used close-micing techniques to prevent leakage from other instruments. Thus the sound we have recorded on the individual tracks are, for the most part, direct sound. During mixdown we are trying to make the sound of a band performing in some kind of room, like a concert hall. In order to obtain this perspective, we must add reverberation to the direct sound. As we add reverberation, the instrument gets further away in the picture. Thus, we can also control how far away the instrument sound by how much additional reverberation we add.

In the past, the reverberation was generated by a device, called an effects processor that was external to the console. Now this is done on an auxiliary channel with a plugin creating these same reverberation effects. The effects plugin needs to be set to a type of reverberation program, like "Concert Hall." The device will then take any signal that is sent to it and generate repeats of the sound that sound like a concert hall.

Important note

Some new engineers put the reverberation plugin on the individual channel and then adjust the amount of reverb using the mix control of the plugin. This is HIGHLY recommended against. Reverberation set this way will reduce the level of signal being mixed with the output and will not allow the proper compression or equalization to be set on the individual channels. That method also required a plugin be set on <u>every</u> channel, taxing the computer system heavily.

The DAW program will have send controls for each channel, and the effects/auxiliary channel will hold the plugin the output of which is mixed with the original channels.

By tuning the individual aux send control on a channel, you can send some of the signal from that channel to the effects device. With these send controls, you can adjust how much reverberation there is on each instrument. On the plugin, there will be an input control which would effects the overall level of all instrument signals that are being sent to the effects plugin, but it is unlikely that you will need to adjust this.

The aux channel output is the equivalent of a console's Echo Return Control. This control adjusts the overall level of the signal from the effects device sent to the stereo outputs of the console. This control will be able to adjust the overall amount of reverberation in the mix. Once again, it is unlikely that you will need to adjust this level, especially at a basic level.

The normal settings for the master input is set as a preset and the fader is about 75% up (0 dB). The send control makes sure that the proper level is received by the effects device. There is usually a meter, in the form of a column of lights, on the effects device which shows the input level to the device. If the effects device is getting too low of a signal, increase the input level, you may then have to reduce the

Equalization

Each DAW channel can be given an equalizer plugin. The equalizer allows the engineer to raise or lower the signals that are in a certain range of frequencies. The equalizer can be used to increase certain harmonic or attack frequencies of instruments to get them clearer sounding. The equalizer can also be used to reduce frequency elements that make instruments sound dull or muddy.

A "Recommended Frequency" chart appears below and can be used as a guide as to which frequencies may need to be increased or decreased in level. Adjusting the levels of signals at certain frequencies is often referred to as *boosting* or *cutting* those frequencies.

Frequency	*Uses*
50Hz	Increase to add more fullness to lowest frequency instruments like foot, toms, and the bass. Reduce to decrease the "boom" of the bass and will increase overtones and the recognition of bass line in the mix. This is most often used on loud bass lines like rock.
100Hz	Increase to add a harder bass sound to lowest frequency instruments. Increase to add fullness to guitars, snare. Increase to add warmth to piano and horns. Reduce to remove boom on guitars & increase clarity.
50Hz	Increase to add more fullness to lowest frequency instruments like foot, toms, and the bass. Reduce to decrease the "boom" of the bass and will increase overtones and the recognition of bass line in the mix. This is most often used on loud bass lines like rock.
100Hz	Increase to add a harder bass sound to lowest frequency instruments. Increase to add fullness to guitars, snare. Increase to add warmth to piano and horns. Reduce to remove boom on guitars & increase clarity.
200Hz	Increase to add fullness to vocals. Increase to add fullness to snare and guitar (harder sound). Reduce to decrease muddiness of vocals or mid-range instruments. Reduce to decrease gong sound of cymbals.
400Hz	Increase to add clarity to bass lines especially when speakers are at low volume. Reduce to decrease "cardboard" sound of lower drums (foot and toms). Reduce to decrease ambiance on cymbals.
800Hz	Increase for clarity and "punch" of bass. Reduce to remove "cheap" sound of guitars.
1.5KHz	Increase for "clarity" and "pluck" of bass. Reduce to remove dullness of guitars.

3KHz	Increase for more "pluck" of bass. Increase for more attack of electric / acoustic guitar. Increase for more attack on low piano parts. Increase for more clarity / hardness on voice. Reduce to increase breathy, soft sound on background vocals. Reduce to disguise out-of-tune vocals / guitars.
5KHz	Increase for vocal presence. Increase low frequency drum attack (foot / toms). Increase for more "finger sound" on bass. Increase attack of piano, acoustic guitar and brightness on guitars (especially rock guitars). Reduce to make background parts more distant. Reduce to soften "thin" guitar.
7KHz	Increase to add attack on low frequency drums (more metallic sound). Increase to add attack to percussion instruments. Increase on dull singer. Increase for more "finger sound" on acoustic bass. Reduce to decrease "s" sound on singers. Increase to add sharpness to synthesizers, rock guitars, acoustic guitar and piano.
10KHz	Increase to brighten vocals. Increase for "light brightness" in acoustic guitar and piano. Increase for hardness on cymbals. Reduce to decrease "s" sound on singers.
15KHz	Increase to brighten vocals (breath sound). Increase to brighten cymbals, string instruments and flutes. Increase to make sampled synthesizer sound more real.
	Recommended Frequencies Chart

Don't use an equalizer just to use an equalizer. If the sound is good already, you don't need an equalizer. Use the equalizer on instruments that are dull, hard to hear or sound thin, once you have tried to get the best sound with regular level adjustments. After adjusting the equalizer on one instrument, listen to it in the mix to see how the mix sounds. You may want to make level changes after you equalize. Remember that the objective is a good mixdown. The individual sound of an instrument has less importance - only the sound of the instrument while it is playing with other instruments has great importance.

The equalizer will often have 4 different sets of controls; one for the bass frequencies; two for the midrange frequencies; one for the high frequencies. Each set of controls will include a *frequency* control. The *frequency* control may be labeled *Hz* (for "hertz") or *kHz* (for thousands of hertz). On some inexpensive boards, the bass and high frequency controls will be adjusted to a certain frequency range and cannot be changed. Adjust the Frequency control according to the recommended guidelines.

In each set of controls, there will be a control for the amount of boost or cut. Equalizers usually allow you to increase boost the frequency range to 4 times as loud or cut it to 1/4 as loud. The *amount* control may be labeled *dB*. Until you are well experienced, keep your boost and cut to about half this amount.

Chapter 4 – More About Microphones

Recommended Method of Obtaining a Mix

The following is an outline of the steps involved in obtaining a good mixdown:

1. Setup

   Turn down all faders to lowest level. Put the console in the Mix status or set Aux channel to Solo Defeat. Ensure the instrument name is recorded on each track. Bring up the Stereo Output to 0 dB, as well as the Echo Return control. Put the speaker/headphone control at about 60%

2. Track Setup

   Adjust the transport in the DAW program to the beginning of the recording or the beginning of the drum track recordings.

3. Build up Basic Levels & Pans

   a. Bring up the foot drum, panned center, until the stereo meters read approximately "-6." You will leave this level alone and mix your other instruments around this level.
   b. Bring up the snare drum, panned center, until the snare sounds as loud as the foot, by ear.
   c. Bring up the high hat, panned slightly right, until it sounds as present in the mix as the snare.
   d. Bring up the rest of the drum faders, panned left & right or as appropriate. Using faders only, adjust the level of the drums and cymbals for the most even sound. Be sure to leave the foot drum fader alone, adjusting other faders instead.
   e. Bring up the bass, panned center, so that the bass and foot are the same level by ear.
   f. Bring up the rhythm instruments, panned left & right (or left right & center) so that each of the rhythm instruments is the same volume as the bass, by ear.

      Note that you can use the solo buttons to make sure that one instrument is as loud as another. Remember, however, to listen more to the mix than to individual instruments.

   g. Bring up any background vocal tracks, evenly panned, so that the background vocals are slightly louder than the rhythm instruments.
   h. At this point the average level of the meters should be peaking to "-2" on the stereo output meters. If the levels are slightly higher, reduce the stereo master slightly.
   i. Add the lead vocal so that it is louder than the mix than the background vocals and the stereo meters peak at around "0".
   j. Add any other lead or accent instruments at about the same level as the background vocals. If there is more than one lead instrument, pan for the best balance.
   k. Look at your master output meters. If necessary, adjust the stereo master fader so that the peaks are at or slightly higher than "0."

4. Add Reverberation.

   Use no or *very little* reverberation on the foot or bass. If there is a piano, use very little reverberation (because there is natural reverberation in the piano box). Use the most reverberation on the drums, rhythm instruments and background vocals to place them further away. Use the same amount of reverberation for all of the drum tracks (except the foot) to get them sounding the same distance away. Use less reverberation on the lead vocal and the lead instruments to get them sounding more "up front."

5. Equalization

   Equalization is added last so that you can hear its effect on the overall mix. Listen to the mix at different volumes. Equalize the instruments that are hard to hear, dull, muddy or thin in the mix. Equalize the bass instruments by boosting harmonics or attack if they are hard to hear at low volumes. Each time you equalize an instrument, make sure that your fader levels don't have to be adjusted.

6. Fader Moves

   Practice any fader moves that will enhance the mix. Sometimes a guitar plays a rhythm part for most of the tune, and then has a solo lead guitar part in the middle of the tune; it may have to be brought up for the solo section of the tune and the panning may need to be adjusted to the center. Due to difficulties in the performance or in the recording, an instrument or vocal may have weak or strong lines - these can be evened out with fader movements. Practice the fade-out with the stereo master fader at the end of the tune, if this is required.

7. Check it by listening back to it at different levels.
8. Record your mix.

Mixing Demonstration

Your instructor will perform a basic mixdown of a tune for you, following the above guidelines.

CHAPTER 7 – RECORDING RAP AND HIP HOP

Fifty years ago you recorded with musicians and singers. Today you record with musicians and singers but also with "sounds in a box," made possible with advancing technology.

The Drum Machine

By the late 1970's Lynn Drum had come out with a studio drum machine that sounded like real drums. When you pressed a drum button, the unit played back a digital recording of a drum or cymbal. Soon producers were using these machines rather than real drummers. Technology had created an alternative method of recording.

By the mid 1980's Rap productions appeared on the market during the summer months. Although this form of performance had been around much longer, it began to get very popular in the mid-80's. With rap you had people perform rhythmic talking over drumbeats. To this you could add a bass line or a new form of music, the looped sample. A sample is a short recording off of a record that is repeated over and over to get a "bed" of music to rap over.

By the mid-80's a cousin of rap music began to get popular: Hip-Hop. Hip Hop is a combination of beats, samples, rap, musicians playing and singers singing, all in a dance-style performance.

The important point is that these styles of music and styles of recordings were advanced by technology making other recording methods possible. Many thought that Rap and Hip-Hop were passing fads that would never reach the mainstream music scene. Many considered this style of production a cheap, uncreative form of recording that required no talent. Today these styles of recording often take more work and more creativity to pull off then merely "having musicians play." The art value is, as always, left to the judgment of the listening public.

The Sounds

As previously stated, the drum machine was and is a device that is designed to program and play back pre-recorded drum hits. Since the sounds of the drums are actually digital recordings, they can sound just like drums or can sound like drums with effects added to the real sounds. The drum sounds can sound like electronically made versions of real drums.

The drum machine often came with drum sounds loaded into computer memory chips. Many drum machines would allow the user to load in additional drum sounds from a memory cartridge or computer disc. These machines sometimes let the operator use some of the sounds from different sources so that the sounds used on the production are a variety of sounds from internal memory and several memory cartridges. The largest drum machines will also let the user sample new drum sounds and load them into memory. These new drum sounds can be stored onto a disc.

Virtual Studio Technology and Plugins

Although still available and used by well-known producers, the drum machine, like other technology before it, has largely been replaced by programs and plugins designed to do the same job at a fraction of the price.

Programming a Beat

There are different types of plugins which mirror different technologies used to create patterns.

Virtual Instruments are added to a track which can also hold MIDI commands. The track is put in record ready and the user

The user can then either record the pattern of drums sounds in real time or draw them in as MIDI notes, or using (if available) a MIDI Drum Editor function.

Pattern Programming Demonstration

Your instructor will show you the programming of a drum pattern. A set of drum sounds will be loaded into the virtual instrument of the instructor's choice. The instructor will then program a simple pattern using the timing correction feature/quantization.

Sampling & Looping

The sampler was a specialized digital recorder. The sampler was used to record a short segment of music. The sample could be recorded from a released record (with permission), from a release that is designed to be sampled or from a recording that is made to be sampled. The sampler records into internal computer memory chips and has a limited capacity of time that can be recorded.

The MPC drum machine could act as a sampling unit, allowing the engineer to program more than just drum sounds.

The first step is to isolate a track or source that a sample will be taken from.

The engineer then edits the sample down until only the desired portion is left, often a portion that can loop smoothly.

To ensure it is edited properly, the sample is played against a metronome. The tempo and the sample are adjusted until it is playing smoothly along with the metronome.

The sample is then "bounced", "rendered" or "consolidated" to create a file with just that portion.

When using it as a part of a beat, the engineer can adjust the tempo or pitch to match the project that is being worked on.

Sampling and Looping Demonstration

Your instructor will sample a cut from an audio file. After the sample has been recorded, he will truncate the sample so that is the exact length of one or two measures. After the sample is prepared, he will set the DAW program to play back the sample in a correct loop along with the earlier programmed drums.

Pre-Cleared Loops

As an alternate to sampling, many producers use loops that are pre-cleared to avoid later legal requirements to get a sample cleared before releasing a song.

There are programs and plugins which are specifically made to allow the easy programming of patterns and loops. These programs include FL Studio by Image-Line, Reason by Propellerhead and Sequel by Steinberg.

Reason and FL studio can be rewired into other DAW programs and used as instruments. Projects in Sequel can be opened and used directly in Cubase Pro, another DAW program.

Recording the Production

In today's modern studio, it is a common technique to obtain or create a beat and then to record vocals against that.

Sometimes the beat used already has compression and the mixing process requires matching the compression characteristic of the beat to the compression added to the vocal.

It is a very common way to record the genre, but it is not the recommended method.

Mixing the Beat and Vocal

If the engineer has the beat and its various parts on separate tracks, along with the vocal and other added parts, standard mixdown technique can be applied and the result will be a professional and clear sound.

CHAPTER 8 - USING DIGITAL SYSTEMS

Digital Audio

Originally the sound energy put out by the instrument was in the form of a Sound Pressure Wave (as discussed in Chapter 1). The vibrating source (vocal cord, guitar string, etc.) causes moving wave of air pressure changes.

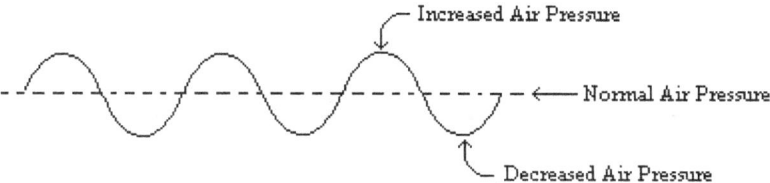

Sound Pressure Wave

Once the microphone picks up the sound pressure wave, it converts it into an alternating push and pull of electrons in a wire (as discussed in Chapter 3). This "wave" of electrical energy is called an audio signal. The audio signal is an <u>analog</u> signal. An analog signal is characterized by continuous changes that match the continuous changes of something else. The audio signal is continuous voltage changes that match the continuous pressure changes of the sound pressure wave.

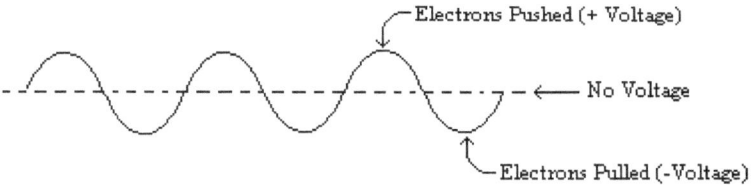

Audio Signal

When the audio signal enters a digital device (such as a digital interface), the audio signal's level is measured several thousand times a second. There is a pulse signal that is generated that represents the measurements. This pulse signal is called a Digital Audio Signal.

The digital audio signal is actually composed of 1's and 0's because computers and digital devices can only recognize the number symbols of 1 and 0. The digital systems can't recognize 2, 3. 4. 5. 6, 7, 8 or 9 like humans can. The digital device makes up for this limitation by stringing sixteen 1's and 0's together to represent one value. Each set of 16 digits, representing one value of audio level, can express 65,536 different possible levels. Out of these 65,000+ possible levels, the system will assign the number that most closely represents the actual level of the audio signal at that instant. This is shown on the next page.

The pulse signal that makes up the digital audio signal comes from the system sending a voltage to represent a 1 and not sending voltage to represent a 0. This process is repeated 15 more times to get one value communicated. The system then repeats the process another 16 times to communicate the next value, etc.

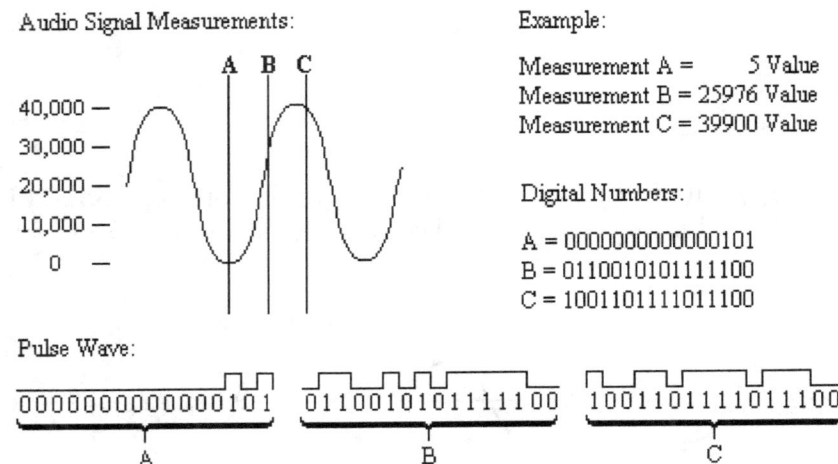

Digital audio signals can run through wires and cause the audio information to be transmitted between devices. If the audio signal has been converted into a digital audio signal by a digital tape recorder, the pulses of the digital audio signal can sent onto another recorder's digital input.

When digital audio is transferred from one device to another, there is no loss, no noise added and no distortion added. Converting a digital audio signal to a (analog) audio signal and then back to a digital audio signal *does* degrade the signal, causing additional noise and distortion. Analog audio signals have a tendency to degrade every time they run through a device; digital audio signals do not tend to degrade by running them through devices.

What is a Digital Interface?

A digital interface takes the audio signals from the microphones and immediately converts them into digital audio signals. This is then sent to the DAW program within the computer.

All of the console functions of routing and mixing are done to the digital audio signals using the interface and program.

Routing: The digital interface's input can be routed to any track of the same format (mono/stereo) in any order in the DAW program, doing the same routing function of any recording console routed to a tape machine.

Mixer #1: The tracks are sent to their locations so they can be routed back to the main outputs.

Mixer #2: This is done within the DAW program by routing the individual tracks and channels to the main output. This is sent to the control room monitors.

Mixer #3: In a similar manner, the DAW program can send signal to another set out outputs which can be sent to the musicians' monitor system, though this is rarely done separately in modern recording.

During the basic recording session, the interface outputs are sent digitally to the computer and recorded there. During mixdown, the digital track outputs are mixed within the computer DAW program without being converted or sent out. Because the signals stay digital audio signals, there is no loss or added noise and distortion.

Chapter 8 – Using Digital Recording Systems

Digital's History and Future

The digital console has been around since the 1980s. For well over a decade. One company, Yamaha, popularized the digital console, starting with their first digital mixing system using the model DMP 7 in the mid-80's. By the late 80's Yamaha released a mid-priced all-digital recording console, the DM-1000 that found great popularity in audio post-production studios doing television and film audio.

In the early 90's several major console manufactures, most notably Neve, started releasing high-end, all-digital consoles in the price range of several hundred thousand dollars. In early 1996, Yamaha released the first all-digital, popularly-priced console that was designed for both recording and mixing, the 02R. Since that time several companies like Mackie and Tascam have released consoles directly competing with the Yamaha 02R. Other companies like RSP have released mid-priced all digital consoles.

All-digital consoles quickly gained ground because:

1. Their use with digital recorders means a cleaner (less distorted and noisy) final product.
2. Cost of the units are close to the cost of analog consoles of comparable size and quality.
3. It was relative easy for manufacturers to include many additional features at very little additional charge. This makes the digital consoles less expensive than the analog console with the same features.

Starting in the late 1990s, with the drop in price of memory and hard drive space, DAW programs started replacing separate digital consoles and recorders.

In modern mixing, almost all recording is done within DAW programs.

Digital Interface Input Levels & Controls

During recording, the digital interface is receiving analog microphone signals. The input controls of a digital console are much the same during recording.

There is sometimes an input switch. In the "Mic" position the microphone signal is received by the interface's input. In the "Line" position a line-level signal may be picked up by the console. Outputs of a synthesizer or analog output of an interface would be examples of line level signals.

The input levels of the digital interface are set using the same procedure used to set the input levels of an analog console. The "Clip" or "overload" light can occasionally flash on loud peaks but should not be solidly lit or flash on every beat of the music. The "Mic Gain" control is set for maximum level without clipping or overload. There is also usually a "pad" switch that would be marked something like "-20 dB." Pushing this pad switch reduces the incoming signal strength to prevent overload. This pad is used when the incoming signal still overloads the console when the input gain control is put at minimum.

Input Selection for Digital Audio Signals

On a digital interface, the channels can also sometimes have a digital input. This allows the channels to receive a digital audio signal from a separate mic preamp.

Setting Channel Controls on a Digital Interface vs Console

One of the first things you will notice when looking at a digital interface is that there are very few knobs, faders and controls on the console. This does not mean that the console has less capability.

There are many different adjustments that have to be made to a channel. Most of this is done within the DAW program. There are the faders for main levels, mute switches to turn the channel on or off, pan controls for left-right-center perspective, routing controls to get the signal to different outputs as necessary, aux send control to add reverb and equalization controls to adjust tone.

On an analog console, these various controls would be in a strip above the channel fader. If the console has 24 inputs, there will be 24 sets of routing controls, 24 pans, etc. On a digital console, you would find a fader, a mute switch and a "select" button. There are two methods that console manufacturers provide to make channel adjustments and settings:

Selected Channel Controls

Many digital consoles, including the Midas 32 Live Sound Mixing Board, have one set of channel controls called the "Selected Channel Controls." When you press the "select" button on a channel, the selected channel controls allow adjustment for that channel. When you press a different "select" button, the controls adjust the newly selected channel.

Function Buttons, Data Entry Knob & Display

An alternate way to adjust channel controls is to use the "Function Buttons." When you push the "Routing" function button, the display screen on the unit changes to a routing screen. You can now move the cursor with the arrow buttons and enter the routing you want with an "enter" button or adjust a "Data Entry" knob to change the display to make the desired routing. If you wanted to adjust the equalization on the selected channel, you would push the "EQ" function button and use the cursor and data entry knob to enter the settings you want. The sequence you push the buttons is important:

1. Push a select button on a channel.
2. Push the function button with the name of the function you want to adjust.
3. Move the cursor over to the parameter or setting you want to adjust.
4. Use the data entry knob to change the display to the setting you want.

Input and Return Channels

The Midas 32 board, when used for living mixing, has all of the functionality of a recording console. When used as an interface, it only uses the selected channel input gain, and the monitor level controls. Even the faders aren't used in interface mode.

Internal Effects

With DAW programs all of the work is done within the program by plugins.

This is a much less expensive operation than the days of analog where you had to buy a console, a recorder, reverb units, external compressors, and all of the wires that hooked them all up.

Setting Levels With Digital

Digital console meters almost always read only "peak" energy and do not read "average" energy of the signal. The digital console, like the digital recorder cannot tolerate any amount of overload. The "0" point on the meters is the maximum level without overload and levels are generally set to cause a "-6" level to provide room for an unexpected higher peak.

CONCLUSION

We hope you have enjoyed this overview of multitrack recording and that it will help your home or studio recording go better.

Obviously there is more to learn.

The Recording Institute of Detroit offers a complete, in depth program of learning called The Recording Techniques Program. In addition to different courses, RID offers different texts as well.

For the beginning engineer, we recommend studying the following in addition to this text:

Pro Audio Specialist Interactive Study Module
(Contact Office for current pricing)

A 61 Lesson web-based module covering all of the basics of audio.

Contact the school office for more information about these offerings at (586) 779-1388.

www.ingramcontent.com/pod-product-compliance
Lightning Source LLC
Chambersburg PA
CBHW080907220526
45466CB00011BA/3500